AN INTRODUCTION TO **SOILS** FOR ENVIRONMENTAL PROFESSIONALS

DUANE L. WINEGARDNER

CRC Press
Taylor & Francis Group
Boca Raton London New York

CRC Press is an imprint of the
Taylor & Francis Group, an **informa** business

T0203617

First published 1995 by Lewis Publishers, Inc.

Published 2019 by CRC Press
Taylor & Francis Group
6000 Broken Sound Parkway NW, Suite 300
Boca Raton, FL 33487-2742

© 1995 by Taylor & Francis Group, LLC
CRC Press is an imprint of Taylor & Francis Group, an Informa business

First issued in paperback 2019

No claim to original U.S. Government works

ISBN-13: 978-0-367-44885-1 (pbk)
ISBN-13: 978-0-87371-939-1 (hbk)

Visit the Taylor & Francis Web site at
http://www.taylorandfrancis.com

and the CRC Press Web site at
http://www.crcpress.com

Figures 3.2 and 3.6 are reprinted with permission of Simon & Schuster from the Macmillan College text, The Nature and Properties of Soils, 10th ed, by Nyle C. Brady.

Library of Congress Card Number 95-35149

Library of Congress Cataloging-in-Publication Data

Winegardner, Duane L.
An introduction to soils for environmental professionals / Duane L. Winegardner.
p. cm.
Includes bibliographical references (p.) and index
ISBN 0-87371-939-5
1. Soil science. 2. Soils. I. Title.
S591 W73 1995
631 43—dc20   95-35149
                                                                 CIP

soli Deo gloria

Preface

Environmental science is a broad spectrum field. Preservation of healthy life on this planet requires the careful management of all the primary natural resources: air, water, and soil. Each field of scientific endeavor has the capacity to contribute understanding to one or more aspects of these resources. As each of us focuses on our individual corner of the universe, we become increasingly aware of the interrelationships between all the various natural (and man-made) environmental parameters. No one portion of our earthly life support system is independent. Constant interplay occurs between the atmosphere, surface water, groundwater, and soil. Plants and animals depend on these complex relationships. Soils are the bridge between mineral matter and life.

During my years as an applied environmental scientist, I have been privileged to work with many talented professionals who understood neighboring fields of study. A limnologist (lake specialist) who also appreciated groundwater taught me many things about aquatic chemistry. A foundation engineer spoke confidently about anaerobic degradation of timber pilings. Environmental professionals continue to expand their knowledge as they mature. Many new professionals, however, are pressured by performance expectations and have little extra time to broaden their perspective.

In this book, the basic principles of each of the major soil science fields are assembled into one volume. It is *not* intended to be a "how to do it" cookbook, but rather as a primer to introduce the reader to the concepts involved. Each chapter focuses on a particular aspect while introducing relationships to other studies.

The preparation of each chapter was assisted by a fellow professional who has special talent and interest in that particular field. Chapter 2, which introduces classifications systems, was reviewed by Demita Winegardner (my daughter-in-law), who is an environmental-civil engineer with Cardinal Environmental. Her assistance with the transfer of basic data between the various classifications schemes is appreciated. Stephen Testa, president of Testa Environmental Services, contributed heavily to Chapter 3 on Soil Mineralogy. He has taught mineralogy for several years with a style which leads the student to understanding.

Chapter 4, Soil Mechanics, was edited by Rick Trapiegnier of the Zia Corporation. Rick is especially talented regarding applied physical soil problems. The chapter on Soil Physics (Chapter 5) was scrutinized by John Barone. John's criticism, based on his theoretical and environmental studies, was very helpful.

Soil Chemistry (Chapter 6), both organic and inorganic, was reviewed by Patrick Francks, who specializes in environmental chemistry. Pat used parts of this chapter as handouts for his class at Oklahoma State University, Oklahoma City Campus; their review was also appreciated.

Lonnie Kennedy, with Deerinwater Environmental Management Services, a specialist in biodegradation, edited Chapter 7 on Microbiology. Lonnie is the author of several published articles on restoration of sites by the use of biological active agents.

Chapter 8, Sampling Techniques, was reviewed by Leonard Billingsley. Because he has evaluated hundreds of soil sampling projects, his criticism was greatly appreciated. Selection of Analytical Procedures (Chapter 9) was supported by the efforts of Linda O'Donnell, who is with Quality Control Laboratory. Her practical experience with both collection and analysis of environmental samples makes her unique to her specialty area.

Agricultural Considerations (Chapter 10) presents the basic concepts of plant and soil interactions. David Klumpp, a respected gardener and friend, provided input to this chapter.

The best conducted scientific project cannot be successful unless the report is meaningful. Chapter 11 discusses Management, Presentation, and Interpretation of Soils Data. Beverly Dowdey, a research specialist with Environmental and Engineering Information Services, reviewed and assisted with this chapter.

The final chapter (12), Case Histories and Applications, was compiled from published articles, public data, and personal experience. I wish to thank *Soils* magazine for their kind permission to abstract articles.

Special thanks are due to the editors at Lewis Publishers, especially Kathy Walters and Vivian Collier, who supported and encouraged me on a regular basis.

Many other friends, associates, and students have encouraged and supported preparation of this text. My special thanks goes to Stephen Testa, Joe Thacker, and Jim Blackwell for preparation of the graphics, and, as always, to my wife, Jane for her tolerance and encouragement during the past 30 months while this book was being prepared.

Duane L. Winegardner

Duane L. Winegardner is a senior hydrogeologist with Cardinal Environmental in Oklahoma City, OK. In this capacity, most of his work involves investigation, evaluations, and engineering designs for remediation of contaminated soil and groundwater. Also, during the past three years he has served as an adjunct professor at the Oklahoma City campus of Oklahoma State University, teaching classes and seminars related to soil and groundwater. Duane earned both his B.S. in Geology (1967) and M.S. in Geology and Hydrology (1971) at the University of Toledo, Toledo, OH. Subsequently, he has achieved registration as a Professional Engineer (Civil) and is licensed in several states. For the past 28 years, his focus has been on applied technology in the construction and environmental industries. His employers have included Toledo Testing Laboratory, St. Johns River Water Management District (FL) Environmental Science and Engineering, OHM Corporation, Engineering Enterprises, and Deerinwater Environmental Management Services.

For the past eight years, his emphasis has been focused on the cleanup of petroleum contamination in soils and groundwater. Many of his remediation designs have been based on new applications of existing technology as well as development of unique processes for specific geological and chemical settings. In 1990, he and Stephen Testa published *Restoration of Petroleum Contaminated Aquifers.*

Duane is a member of the American Society for Testing and Materials, the Association of Ground Water Scientists and Engineers, and the Oklahoma Society of Environmental Professionals. He continues to write professional papers and routinely participates in educational programs.

Contents

x

List of Figures

List of Tables

## CHAPTER 1

The Fundamental Concept of Soil

From a purely pragmatic viewpoint, soil can be considered everything that is included in the superficial covering of the earth's land area. Based on environmental perspectives, soil is an aggregate of unconsolidated mineral and organic particles produced by the combined physical, chemical, and biological processes of water, wind, and life activity.

In common usage, the definition of soil varies according to the user. For agriculturists (and related scientists), the important factors in soil focus on the upper few feet of the soil column, which are important to plant growth. Civil engineers consider soil a structural material with definable physical and chemical properties that can be manipulated (or tolerated) for construction purposes.

Microbiologists are interested in the interaction of microbes in the life cycle of the soil. Soil chemists study the detailed chemical reactions that result from the continuously changing underground laboratory. Questions asked by geoscientists include: Where did it originate?, How did it get here?, What is the next stage of its development?, and Of what is it composed?

The science of soil physics, which evolved from agricultural studies, considers the mechanical functions of soil such as fluid flow, interparticle relationships, and the complex interaction between the soil atmosphere, water, mineral surfaces, and organic matter.

Scientists in the individual disciplines of soil science identified in the preceding paragraphs tend to focus their efforts in specific directions toward the solution of well-defined problems. The environmental specialist is expected to be a "jack-of-all-trades" or (in medical terms) a "general practitioner." Environmental science recognizes that all terrestrial life depends on soil for its existence. It is important that environmental specialists have at least a deep appreciation of the complex interactions of all of the various happenings occurring within the "soil sphere" (no pun intended).

The chemical composition and physical structure of the soil at any given location is determined by: (a) the type of geological material from which it originated, (b) the vegetative cover, (c) length of time that the soil has been weathered, (d) topography, and (e) the artificial changes caused by human activities. Land surfaces almost everywhere are covered by this unconsolidated debris (soil) sometimes called the *regolith*. This blanket above the bedrock may be very thin, or it may extend to depths of hundreds of feet. Its physical and chemical composition may vary, not only horizontally, but vertically, and its geological origin is not always the same, even within the local area. Some soils are formed from the bedrock (or other geological material) which immediately underlies it. Other soils develop as the result of transported materials being deposited in their current location by the action of water, ice, or wind.

Regardless of the origin, most soils consist of four basic components: mineral matter, water, air, and organic matter. These materials are present in a fine state of subdivision (individual particles) and intimately mixed. In fact, the mixture is so completely blended that separation is difficult. The more solid part of the soil, and naturally the most noticeable, is composed of mineral fragments in various stages of decomposition and disintegration. A variable amount of organic matter, depending on the horizon observed, is blended with inorganic substances. Normally, the largest amount of organic matter is in the surface layer.

The mineral material has its genesis in the parent material. Some soil minerals persist almost intact, while others are quickly transformed into new minerals in response to the current soil environment. Various sizes of mineral particles occur, ranging from coarse sand (2 mm in diameter) to finely divided clay particles (< 0.002 mm in diameter).

Organic matter represents the accumulation of plant and animal residues (generally in an active state of decay), especially in shallow horizons. Most organic matter is distributed in finely divided clay-sized particles. The organic content of soil ranges from less than 0.05% to greater than 80% (in highly organic soil such as peat), but is most often found between 2 and 5%.

All soil, even the most dense, has significant pore volume between soil grains. Soil pores are variable as to continuity, local dimensions, and total volume. These void spaces may be filled with air or water. The proportion of water-filled or air-filled pores is dependent upon the character of the soil, the conditions of formation, and when the last water was added to the system. Soft clay may have void spaces of over 60% by volume, while a well-compacted, uniformly sized sand can have pore volumes in the 25% range.

Water in the soil structure can be either transitory or fixed (at least temporarily). The part of the water that is retained is held with varying degrees of tenacity. Surface contact between the water and soil particles is the key to many chemical and physical reactions. Soil water is never without solutes (dissolved components) as its solvent properties, along with varying acidic and oxidation potential, causing it to be a major player in the dynamic nature of soil.

The general components of the soil atmosphere are nitrogen, oxygen, and carbon dioxide. Nitrogen is relatively inert unless fixed by soil bacteria. Free oxygen is present unless it has reacted with mineral matter or organic matter (especially living tissue).

Carbon dioxide can result from respiration of microbes, or abiotic chemical reactions. The concentration of carbon dioxide in the soil air is usually greater than the outside atmosphere. The proportion of oxygen, nitrogen, and carbon dioxide not only varies inversely with the amount of water present in the soil, but is extremely variable as to the ratio of the gases present.

Soils have significant variations in appearance, fertility, and chemical characteristics, depending on the mineral and plant materials from which they were formed and continue to be transformed. The soil realm has been compared to an elaborate chemical laboratory where a large number of reactions occur simultaneously. A few of the reactions are relatively simple and well understood, while the vast majority are not yet completely explained. The reactions range from simple solution and substitution to complex biologically mitigated multistep processes. Many reactions depend on the participation of water, mineral, and biological factors in a dynamic setting.

Soil, then, is a dynamic environment; almost a living structure. Continuous processes (although relatively slow) are active in even the most remote settings. Soil has been called "the bridge between life and the mineral world." All life owes its existence to a few elements that must be ultimately derived from the earth's crust. After weathering and other processes create soil, plants (including microbes) perform the intermediary role of assimilating these necessary elements, making them available to animals and humans.

All mineral energy sources on earth, as we know it, come from plants that have grown in the soil while obtaining their energy from the sun. Most "natural" materials used by man are derived in some way from the soil.

As concern for the environment increases, it is comforting to recognize that soil, when properly used, can offer an unlimited potential for disposal and recycling of waste materials. Knowledge of physical, chemical, and biological reactions is more important to us today than ever before.

CHAPTER **2**

Classification Systems

2.1 INTRODUCTION

The ability to describe a soil with sufficient clarity so that a fellow observer can recognize the soil is an age-old problem. The level of detail required is usually based primarily on the purpose of the description. If, for example, a gardener is writing to a newspaper horticulture columnist, a sufficient description for a soil might be "sandy loam." That level of detail, along with the geographic location, may enable the columnist to conclude that the garden would benefit from the addition of lime, water, and fertilizer to improve green bean production.

On the other hand, an environmental geologist making a detailed investigation of a potential hazardous waste disposal site should (ideally) invest a considerable amount of time and effort describing the minute details of the soil column. The appropriate level of detail in this case will most assuredly include visual observations of structure, texture, color, and apparent soil type, plus laboratory analysis of particle size distribution, and a host of chemical components.

A common practice of inexperienced field professionals is the assumption of values for these parameters based on average characteristics reported in the literature. The probability of failure of a remediation design based on assumed values is great.

Classification of soils should be based on the needs of the user. If insufficient detail or accuracy is presented, the user must either reinvestigate the site, or design the final use with enough safety factors to compensate for weak data. At the other extreme is the scholarly approach that generates more data than the project demands, or the client is willing to afford.

Several of the most common systems of soil classification are presented in this chapter. While each system was developed for a specific purpose, some data included in each can be transferred to other uses. The focus of the discussion in this chapter is oriented to application needs of environmental specialists.

2.2 PARTICLE SIZE ANALYSIS

For environmental purposes, a soil is defined as virtually every type of non-cemented or partially cemented inorganic material found in the subsurface. This wide definition often includes the entire column that can be drilled by a drilling rig equipped with hollow stem augers. Often, weathered shale, sandstone, and other semiconsolidated materials are considered soil for environmental and construction purposes. While not a sophisticated definition, it has a practical value because it defines material that behaves like soil and can be excavated with common earthmoving equipment.

Most subsurface materials that cannot be augured are sufficiently brittle to contain natural fractures. Fluids (gas or water) migrating preferentially through fractures have limited contact with most of the soil grains, and thus have very few opportunities to interact with each grain.

Soil is composed of two basic components: solid matter and pore fluids (water and soil atmosphere). Since pore fluids are usually relatively minor components, they are considered secondary as far as classification is concerned. Solid particles consist of mineral grains or organic matter of various sizes and shapes, occurring in a variety of organizational forms.

Solid particles can be divided into various components based on their size and shape (i.e., sand, silt, or clay), each of which interacts with water to give the soil its individual physical characteristics. The following sections discuss the procedures used to describe the physical characteristics in a scientific manner.

2.3 PARTICLE SIZE DISTRIBUTION

Probably the most basic tool of soil description is that of the size range of mineral components. A casual observer may indicate that a soil is "sand," "clay," or "silt" (or even boulders or gravel), with only a general regard for the definition of each term.

Definition of the range of particle sizes included in each group has been a topic of debate between soil specialists since soils have been classified. Several ranges are commonly in use at the current time. Table 2.1 presents the most widely accepted particle size standards. For most practical environmental purposes, selection of any one standard is acceptable as long as it is clearly defined and employed consistently throughout the project.

The procedures used to separate soil grains by size are determined by the nature of the soil. Granular soils (boulders, gravel, and sand) have grains that are easily seen by the naked eye or handheld magnifying glass. When dried, the grains of these "coarse grained soils" have little tendency to cling together. Mechanical sieves may be used to separate individual grains into separate size ranges. Soils that are "cohesive" (silt and clay), tend to form hard, strong lumps when dry. Sieves with openings small enough to separate particles as small as silt and clay cannot be easily manufactured.

Table 2.1 Grain Size Limits by Various Systems

Grain Size (mm)	Sieve Size	Unified Soils Classification System	American Association of State Highway and Transportation Officials	U.S. Department of Agriculture	Massachusetts Institute of Technology
100	3"	Cobble			
	3/4"	Coarse Gravel	Gravel	Gravel	Gravel
10	1/2"	Fine Gravel			
	4	Coarse Gravel			
	10				
1.0	20	Medium Sand	Sand	Sand	Sand
	40				
	60	Fine Sand			
0.1	140				
	200				
	270		Silt	Silt	Silt
0.01		Fines Silt or Clay			
0.001			Clay	Clay	Clay

Table 2.2   Standard Sieve Sizes

U.S. Sieve No.	Tyler Mesh No.	Millimeters	Inches
4	4	4.7	0.185
6	6	3.33	0.131
8	8	2.36	0.093
10	9	2.0	0.078
13	10	1.65	0.065
16	14	1.17	0.046
20	20	0.833	0.033
30	28	0.589	0.023
40	35	0.417	0.016
50	48	0.295	0.012
60	60	0.25	0.01
70	65	0.208	0.008
80	70	0.177	0.007
100	100	0.149	0.006
130	150	0.104	0.004
140	170	0.088	0.0035
200	200	0.074	0.0029
400	400	0.038	0.0015

Granular soils are separated by sieve analysis that involves selecting a set of sieves with openings which encompass the size range of the soil sample. Table 2.2 presents common sieve size openings. The sieve stack is assembled vertically, with the largest opening sieve on the top and the smallest at the bottom. A catch pan at the base retains the particles that pass all of the sieves. After a weighed quantity of oven-dry soil has been placed in the top sieve, a lid is attached to the top, and the entire column of sieves is inserted into a mechanical shaker (see Figure 2.1). The shaker is operated until the soil particles have reached the sieve where they do not pass (usually five minutes).

Soil retained on each sieve is weighed. The cumulative weight percentage retained on each successive screen is plotted on the vertical axis of a graph, with the sieve openings plotted on the horizontal axis (see Figure 2.2).

The particle size distribution of clay and silt materials is determined by observing the settling time of the particles in water as described by Stoke's law. The principle of operation assumes that the density of the soil particles is within a narrow range. Stoke's law states that the velocity of settlement is proportional to the square of the particle's radius in the formula:

$$v = k\, r^2$$

where:

v = velocity
r = particle radius
k = a proportionality constant

Figure 2.1 Mechanical sieve shaker (photo provided by the Gilson Company).

Measurement of the velocity of settling is made in the laboratory by recording the density of a soil-water mixture over time. As each size of particle settles, the density of the mixture decreases. The density of the soil-water mixture is measured by a hydrometer (see Figure 2.3). Detailed procedures are presented in ASTM D-422.

Figures 2.4 and 2.4 (b) are examples of the graphical presentation of particle size distribution for two separate soil samples. Figure 2.4 (a) represents a well graded (wide range of particle sizes) gravel and Figure 2.4 (b) was developed from a combination of hydrometer and sieve analysis of a poorly graded (limited size range) sand sample. Each result is plotted on graph paper

Sieve Analysis Results

Sieve	Inches	% Passing	% Retained
# 10	0.078	99.1	0.9
# 20	0.033	96.7	3.3
# 40	0.016	78.2	21.8
# 100	0.006	11.3	88.7
# 200	0.003	1.9	98.1

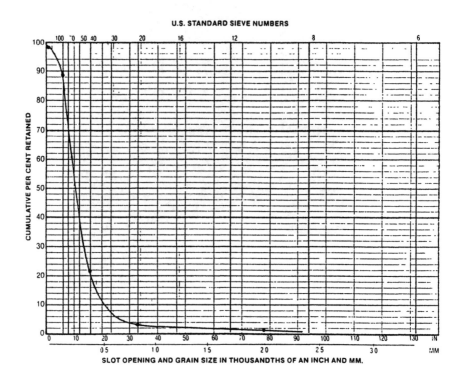

Figure 2.2　Graphical plot of sieve analysis of sandy soil.

presenting the particle size in millimeters and standard sieve sizes. The two vertical sides of the graph present the data as percent passing and percent retained,
respectively. Also, the general size ranges (sand, gravel, etc.) are shown for
convenience.

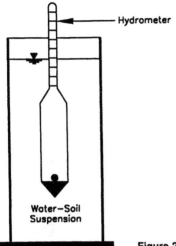

Figure 2.3   Hydrometer test in progress.

The general slope of a size distribution curve describes the range of particle sizes present in the sample. Samples that contain a wide range of sizes have slopes that are less steep, while uniform one-size samples have slopes which are almost vertical. The slope of the curve is characterized as the *coefficient of uniformity* (C_u) which is determined by:

$$C_u = D_{60}/D_{10}$$

where:

D_{60} = particle diameter at 60% passing
D_{10} = particle diameter at 10% passing

The D_{10} is also known as the *effective particle size*. Particles of this diameter control most of the physical characteristics of the soil sample (i.e., permeability, cohesion, and density) because they fill the smallest void spaces between the larger particles.

2.4   SOIL CONSISTENCY

The physical properties of most fine grained (smaller than No. 40 sieve) soils are controlled by water content. If clay soil contains a certain amount of water it may exhibit a plastic tendency, but if additional water is added it may approach a liquid state. In 1911, A. Atterberg developed a series of tests to define the states of fine grained soil (liquid and plastic) based on water content.

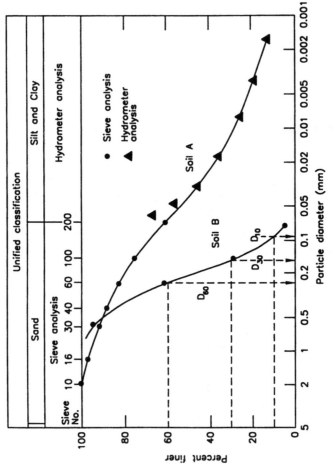

Figure 2.4 Particle-size distribution curve. (a) Soil A is a well-graded soil (wide range of particle sizes). (b) Soil B is a poorly graded soil (narrow range of particle sizes).

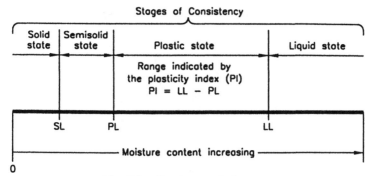

Fig. 2.5. Consistency limits

The liquid limit, LL, is defined as that water content expressed as a percentage of the dry weight of soil at which the soil first shows a small but definate shear strength as the water content is reduced. Conversely, with increasing moisture, it is that water content at which the soil mass just starts to become fluid under the influence of standard shocks.

Figure 2.5 Consistency limits.

Atterberg observed that fine grained soils can exist in four distinct states: solid, semisolid, plastic, and liquid. Each state of consistency is dependent on the moisture content of that soil. Figure 2.5 describes the relationship between these states. While the division between these states is somewhat relative, Atterberg developed a series of standard testing procedures to provide a numerical basis for comparison.

The practical significance of Atterberg's tests is to determine the stability of silt and clay at a range of moisture contents. If a natural soil is sufficiently wet to be plastic, it can easily deform when an extra weight load is applied. When dried, the same soil may exhibit great bearing strength. However, after a period of wet conditions, the soil may approach a liquid state and lose almost all of its load supporting ability. It is very important to understand the potential reactions of a silt or clay soil before constructing a foundation or making an excavation.

Liquid Limit (LL) is defined as the water content of a soil at the point where the soil exhibits a slight shear strength. It is measured by an apparatus (see Figure 2.6) designed to cause a slight shock, which causes closure of a groove made in a soil sample. When the moisture content is such that a one-inch-long groove is closed by 25 measured shocks, the liquid limit has been defined.

As the water content is decreased below the liquid limit, the soil enters the plastic state. The lower end of plasticity is the *Plastic Limit* (PL), which is defined as the moisture content at which a soil may be rolled out by hand into 1/8-inch diameter threads.

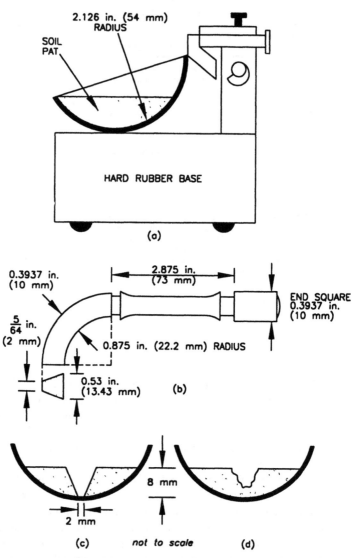

Figure 2.6   Liquid limit test. (a) liquid limit device, (b) grooving tool, (c) soil pot before test, (d) soil pot after test.

The difference between the Liquid Limit and the Plastic Limit is known as the *Plasticity Index* (PI). The plasticity index indicates the range of moisture content through which the soil exhibits plastic properties. This range is important for estimation of structural stability. Silts have low or no plastic indices, while clays have higher values.

As the moisture content decreases below the plastic limit, the soil becomes a solid which can only be deformed with increased force; however, cracks develop as drying continues. When the moisture content is sufficiently low, such that shrinkage stops and cracks no longer develop, the soil has reached the *Shrinkage Limit* (SL).

2.5 PARTICLE SHAPE

The shape of individual particles has important influence on the physical properties of soil. Round particles tend to be more easily separated than flat particles. Soils that are predominately silt (loess) are plate-like and are often able to maintain almost vertical walls in excavations. The low permeability of clay soils is partly due to the plate-like shape of clay mineral grains. The following shapes are the most common:

Bulky or equidimensional grains. These shapes may include rounded, subrounded angular, and subangular, as described on Figure 2.7. Coarse-grained soil components tend to be included in this category, which usually consists of quartz and feldspar.

Flaky or plate-like grains. These are common shapes in the finer grain size fraction. Most clay minerals and mica occur as flat plate-shaped grains.

Elongated grains and fibers. These shapes are encountered in soil containing the clay mineral halloysite, asbestos, certain varieties of volcanic ash or organic soils (peat).

2.6 STRUCTURE

Soil structure is a product of a unique set of interactions of physical and chemical influences that is characteristic of the local soil environment. This orientation of grains affects the retention and migration of liquids, as well as the passage of gases through the soil. Structure also affects the physical-bearing strength, and erodibility.

Soil structure describes the geometric orientation of the soil grains with respect to each other. Factors influencing structure include: grain shape, size, mineralogical content, and water content. Different structures develop in cohesive and noncohesive soils.

The basic structures which develop in cohesive (clay and silt) soils are dependent on the attractive forces acting between clay particles and water. Most clay particles have a predominantly negative charge on their flat surface and some positive charge on the edges of the plate-like crystals. Negatively-charged areas attract cations, and positive charges attract anions. However, when the soil grains can move (as when they are suspended in water), some of the particles are able to balance the charges by orienting themselves into edge-to-face contact (see Figure 2.8).

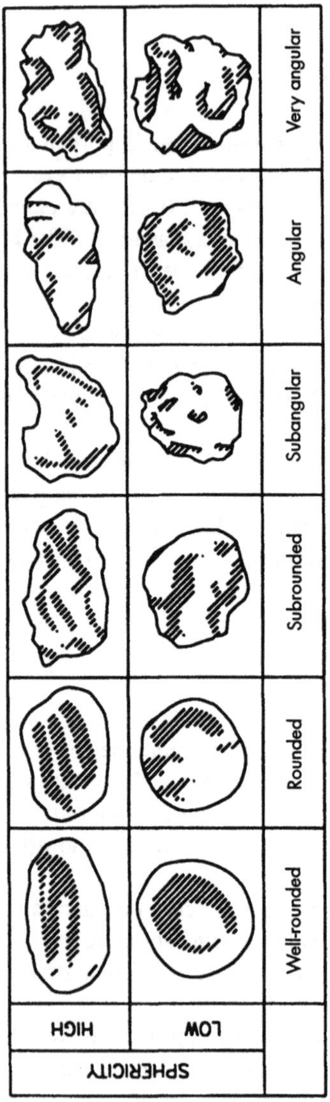

Figure 2.7   Soil grain shapes.

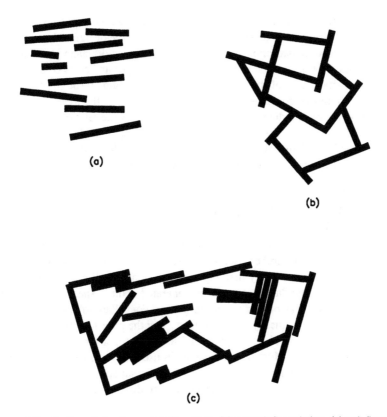

(a)

(b)

(c)

Figure 2.8 Sediment structures. (a) dispersion, (b) nonsalt flocculation, (c) salt flocculation. (Adapted from Lambe, 1958.)

When a sufficient number of the edge-to-face particles become assembled
together, the force of gravity eventually exceeds the repulsive forces between
the grains, and flocculation occurs. If a salt is added to clay suspended in water,
the positive ions of the salt attach to negatively-charged surfaces, reducing the
tendency for repulsion between particles, and often result in flocculation. The
structure of clay soils resulting from flocculation retains a high percentage of
void space.

Soil structures existing in cohesionless (sand and gravel) soils are based on
both single-grained and "honeycombed" organizations. Single-grained structures
are the result of the orientation of particles into intimate contact with the surrounding particles. The compactness of this type of structure is dependent on the
shape and size distribution of the particles, in addition to the packing pattern.

Honeycombed structures sometimes occur in soils that contain large percentages of fine sand and silt. Under some depositional settings, the sand and
silt will be formed into arch-like structures with large open voids. These arches
are stable under light weightloading, but tend to collapse when subjected to vi-
bratory shock or overloading.

Agriculturalists usually want soil to be loose and highly porous. Engineers evaluate the structure to determine its suitability for the design at hand, whether it involves road construction, building foundations, or drainage structures. Environmental scientists often must determine the shapes and sizes of pore spaces, effective surface area of the soil, and flow paths of fluids through the soil. Soil structure should be considered as an important aspect of the local environment.

Soil structure is indicated in field investigations by the relationships between: in-place density, soil type, moisture content dry density, and visual description (i.e., crumbly or firm). While not precise numerical data, field observations of structure are essential elements of a soil description.

2.7   CLASSIFICATION SYSTEMS

The main purpose for classification systems is to describe a soil with sufficient clarity so that another user would recognize it and utilize data gathered about that soil. It is unlikely that any applied scientist will have need for every element of scientific data that could be generated about a specific soil, and it is not economically feasible for an environmental consultant to collect all original data with the precise quality and complexity that the pure scientist would desire.

During the historical development of soil science, several types of classification have evolved. Most of the systems in use today are based on criteria of texture, use, and detailed soil taxonomy for universal application. Each of these systems produces some (or much) data that are directly applicable to environmental practitioners.

2.7.1   Textural Classification

The most fundamental classification system utilizes particle size distribution. Many characteristics of a soil are dependent on the sizes of the grains that compose it. Usually the soil is named after the size of the principal component; sand, silt, and clay are typical terms. Since almost all soils are mixtures of particle sizes, modifiers are necessary. Silty sand, sandy clay, and silty clay are typical examples of descriptive terms.

While the general terms: sand, silt, and clay, are widely accepted and readily recognizable to the casual user, the precise size definitions are not universal. Table 2.1 presents the most common size distributions. For purposes of environmental studies, the selection of a definition is not a critical matter as long as the usage is consistent. Throughout this text, the textural classifications developed by the U.S. Department of Agriculture will be used.

Figure 2.9 is a trilinear diagram that has been adapted for soil textural classification. This particular diagram is intended for use with soils that pass the No. 10 sieve. The three apexes represent sand, silt, and clay. Along each side is a fractional percent scale.

Application of this diagram is described in the following example.

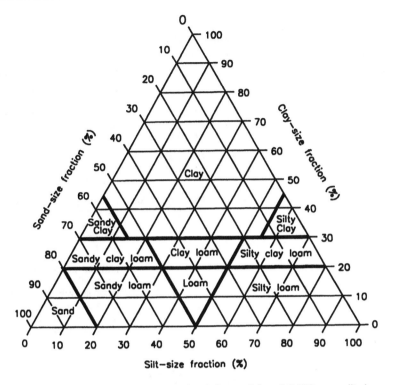

Figure 2.9 Textural classification chart. Sand-size particles, 2-0.074 mm; silt-size particles, 0.074-0.002 mm; and clay-size particles, less than 0.002 mm.

Particle size analysis by a laboratory indicates that this soil is composed of: 48% sand, 42% silt, and 10% clay (all by weight percentage). When the data are plotted on the diagram (as shown by the arrows), the soil is found to be a loam.

Another particle-size-based system that is occasionally encountered is the Burmister Identification System (BS). This description system involves rather precise soil fraction descriptions using specific terminology and symbols. The definitions and symbols used in this system are presented in Table 2.3. The BS uses some capital letters as symbols for the components, with lower case letters to indicate proportionality or gradation. The proportionality terms "and," "some," "little," and "trace" are sometimes incorporated into written descriptions in other classification systems. The BS requires considerable laboratory and field practice before accurate identifications can be made. Classification by soil texture is a useful tool for general soil description. The general nature of the results infers that the soil has characteristics within typical ranges. It does not, however, provide precise data about key parameters such as texture, porosity, permeability, mineral content, organic fraction, density, or chemical parameters, all of which are important to environmental specialists.

Table 2.3   Burmister System Terms

1. Definition of Soil Components and Fractions

Material	Symbol	Fraction	Sieve Size	Definition
Boulders	Bldr	—	9″	Material retained on 9″ sieve
Cobbles	Cbl	—	3″ to 9″	Material passing the 9″ sieve
Gravel	G	Coarse (c) Medium (m) Fine (f)	1″ to 3″ 3/8″ to 1″ No.10 to 3/8″	Material passing the 3″ sieve and retained on the No.10 sieve.
Sand	S	Coarse (c) Medium (m)	No.30 to No.10 No.200 to No.30	Material passing the No.10 sieve and retained on the No.200 sieve.
Silt	$	—	Passing No.200 (0.075)	Material passing the No.200 sieve that is no-plastic in character and exhibits little or no strength when air dried.
Organic Silt	(OS)	—	—	Material passing the No.200 sieve which exhibits plastic properties within a certain range of moisture content, and has fine granular and organic characteristics.

		Plasticity	Plasticity Index	
Clayey Silt	Cy$	Slight (S)	1 to 5	Clay Soil:
Silt & Clay	$&C	Low (L)	5 to 10	Material passing the No.200 sieve which can be made to exhibit plasticity and clay qualities within a certain moisture content, and which exhibits considerable strength when air dried.
Clay & Silt	C&$	Medium (M)	10 to 20	
Silty Clay	$yC	High (H)	20 to 40	
Clay	C	Very High (VH)	40 plus	

2. Definition of Component Proportions

Component	Written	Proportions[a]	Symbol[a]	Percentage Range by Weight
Principal	CAPITALS	—	—	50% or more
Minor	Lower case	and	a.	35–50%
		some	s.	20–35%
		little	l.	10–20%
		trace	t.	1–10%

[a]Minus sign (−) = lower limit, Plus sign (+) = upper limit, No sign = middle range. Signs are used with proportions words or symbols to indicate lower or upper extreme of percentage range.

2.7.2   Classification by Use

When the primary purpose of soil classification is focused on application, the inclusion of specific characteristic parameters is an important improvement. Two related classification systems have been developed to assist engineers with evaluation and design of civil engineering facilities. Both the Unified Classification and American Association of State Highway and Transportation Officials (AASHTO) systems include the properties of both grain size distribution and soil texture.

Unified Soil Classification System (USCS)

This system was developed during World War II to assist in the development of standardized design practices of military facilities. After several revisions, it has found wide acceptance among engineers. The complete procedure, described in ASTM D-2487, is summarized on Table 2.4. Appendix A presents a field procedure for soil classification according to this procedure. The USCS classifies soil into two broad categories:

Coarse grained soils: Coarse-grained soils are subdivided into gravels and sands, depending on the major coarse-grained constituent. For purposes of identification, coarse-grained soils are classified as gravels (G) if the greater percentage of the coarse fraction (retained by the No. 200 sieve) is larger than that retained by the No. 4 sieve, and as sand (S) if the greater portion of the coarse fraction is finer than the No. 4 sieve. Both the gravel (G) and the sand (S) groups are further subdivided on the basis of uniformity of grading and percentage and plasticity of the fine fraction (passing the No. 200 sieve).

Modifiers to coarse-grained descriptions are:

W - well graded (broad size distribution)
P - poorly graded (narrow range of size distribution)

GW and SW soil groups include well graded gravels and sands with less than 5% nonplastic fines. GP and SP soil groups are poorly graded (or "skip graded") soils with less than 5% fines.

Gravels and sands containing more than 12% fines may be classified either GM, SM, GC, or SC, depending on the plasticity of the minus (passing) No. 40 soil fraction. Silty gravels (GM) and silty sands (SM) have liquid limits and plasticity indices that plot below the A-line of the plasticity chart shown as Figure 2.10. Clayey gravels (GC) and clayey sand (SC) will have liquid limits and plasticity indices that plot above the A-line. (Grading is not a factor when the percentage of fines is greater than 12%.)

A dual classification symbol (i.e., GC-GP) is used when the minus No. 200 fraction (i.e., passing the No. 200) is between 5 and 12% and the soil has characteristics intermediate between the two groups.

Table 2.4 Unified Soil Classification System

Major Divisions		Group Symbols	Typical Names	Field Identification Procedures (excluding particles larger than 3 inches and basing fractions on estimated weights)			Information Required for Describing Soils
1	2	3	4	5			6
Coarse-grained Soils — More than half of material is larger than No. 200 sieve size. The No. 200 sieve size is about the smallest particle visible to the naked eye. **Gravels** — More than half of coarse fraction is larger than No. 4 sieve size	**Clean Gravels** (Little or no fines)	GW	Well-graded gravels, gravel-sand mixtures, little or no fines	Wide range in grain sizes and substantial amounts of all intermediate particle sizes			For undisturbed soils add information on stratification, degree of compactness, cementation, moisture conditions, and drainage characteristics.
		GP	Poorly graded gravels or gravel-sand mixtures, little or no fines	Predominantly one size or a range of sizes with some intermediate sizes missing			Give typical name: indicate approximate percentage of sand and gravel, maximum size: angularity, surface condition, and hardness of the coarse grains; local or geologic name and other pertinent descriptive information; and symbol in parentheses
	Gravels with Fines (Appreciable amount of fines)	GM	Silty gravels, gravel-sand-silt mixture	Nonplastic fines or fines with low plasticity (for identification procedures see ML below)			
		GC	Clayey gravels, gravel-sand-clay mixtures	Plastic fines (for identification see CL below)			Example: Silty sand, gravelly; about 20% hard, angular gravel particles 1/2 in. maximum size; rounded and subangular sand grains, coarse to fine; about 15% nonplastic fines with low dry strength; well compacted and moist in place; alluvial sand. (SM)
Sands — More than half of coarse fraction is smaller than No. 4 sieve size (For visual classification, the 1/4-in size may be used as equivalent to the No. 4 sieve size)	**Clean Sands** (Little or no fines)	SW	Well-graded sands, gravelly sands, little or no fines	Wide range in grain size and substantial amounts of all intermediate particle sizes			
		SP	Poorly graded sands or gravelly sands, little or no fines	Predominantly one size or a range of sizes with some intermediate sizes missing			
	Sands with Fines (Appreciable amount of fines)	SM	Silty sands, sand-silt mixtures	Nonplastic fines or fines with low plasticity (for identification procedures see ML below)			
		SC	Clayey sands, sand-clay mixtures	Plastic fines (for identification procedures see CL below)			

Field Identification Procedures on Fraction Smaller than No. 40 Sieve Size

Major Divisions	Group Symbols	Typical Names	Dry Strength (Crushing characteristics)	Dilatancy (Reaction to shaking)	Toughness (Consistency near PL)	Information Required for Describing Soils
Fine-grained Soils — More than half of material is smaller than No. 200 sieve size. Silts and Clays: Liquid limit is less than 50	ML	Inorganic silts and very fine sands, rock flour, silty or clayey fine sands or clayey silts with slight plasticity	None to slight	Quick to slow	None	For undisturbed soils add information on structure, stratification, consistency in undisturbed and remolded states, moisture and drainage conditions.
	CL	Inorganic clays of low to medium plasticity, gravelly clays, sandy clays, silty clays, lean clays	Medium to high	None to very slow	Medium	Give typical name: indicate degree and character of plasticity; amount and maximum size of coarse grains; color in wet condition; odor, if any; local or geologic name and other pertinent descriptive information; and symbol in parentheses
	OL	Organic silts and organic silty clays of low plasticity	Slight to medium	Slow	Slight	
Silts and Clays: Liquid limit is greater than 50	MH	Inorganic silts, micaceous or diatomaceous fine sandy or silty soils, elastic silts	Slight to medium	Slow to none	Slight to medium	Example: Clayey silt, brown, slightly plastic, small percentage of fine sand, numerous vertical root holes; firm and dry in place; loess (ML)
	CH	Inorganic clays of high plasticity, fat clays	High to very high	None	High	
	OH	Organic clays and silts of medium to high plasticity	Medium to high	None to very slow	Slight to medium	
Highly Organic Soils	Pt	Peat and other highly organic soils	Readily identified by color, odor, spongy feel, and frequently by fibrous texture			

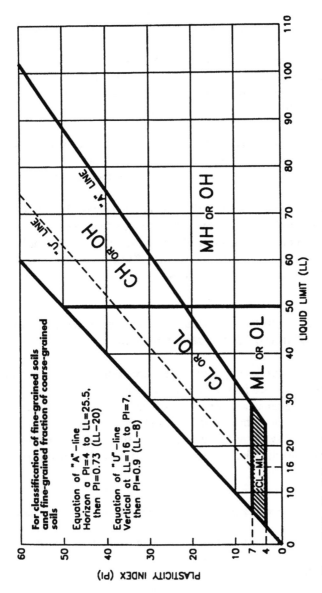

Figure 2.10 Plasticity chart.

Fine grained soils: Fine grained soils are subdivided into silts and clays (depending on their plasticity), and highly organic soils. To distinguish between silts and clays, the plasticity index vs. the liquid limit is plotted as shown on Figure 2.11. Among the inorganic materials, the clays plot above the A-line and silts below the A-line. Silts and clays are further subdivided into low (L) and high (H) plasticity, depending on whether the liquid limit is less than 50% (L) or greater than 50% (H).

Field techniques for differentiating between silts and clays include manual tests for dry strength, dilatancy, and toughness (see Appendix A). These tests are also described extensively in the references. Accurate identification of silts and clays is primarily a matter of experience and conscientious comparison to laboratory test results.

If a fine grained soil appears to contain sufficient organic matter to affect its properties, additional laboratory testing is recommended. Designation of organic silt is "O," which is modified by either L or H to signify the degree of plasticity.

Highly organic soils are combined into a single classification with the symbol "Pt," and are characterized by high organic content commonly consisting of: leaves, grass, branches, and other fibrous matter. These soils usually have high compressibility and relatively low strength. Typical samples of highly organic soils are peat, humus, and swamp soils.

American Association of State Highway and Transportation Officials (AASHTO) Classification System

This system was developed to assist in the design and construction of highways. Soils are divided into seven major groups, A-1 through A-7. Groups A-1, A-2, and A-3 are granular materials which have 35% or less material that passes the No. 200 sieve. If more than 35% of the material passes the No. 200 sieve, the soils are mostly silt and clay, and are classified into groups A-4, A-5, A-6, and A-7. The classification system is based on the following criteria:

(a) Grain Size:
   Gravel: material that passes the 75 mm (3 inch) sieve and is retained on the No. 10 sieve.
   Sand: material that passes the No. 10 sieve and is retained on the No. 200 sieve.
   Silt and Clay: pass the No. 200 sieve.
(b) Plasticity:
   Soils are silty when they have a plasticity index (PI) of 10 or less. When the PI is 11 or greater, the soil is classified as clayey.
(c) Boulders:
   Particles larger than 75 mm (3 inch) are excluded from the portion of the soil that is used for classification, but are recorded.

General Classification	Granular materials (35% or less of total sample passing No. 200)						
	A-1		A-3	A-2-4	A-2-5	A-2-6	A-2-7
Group Classification	A-1-a	A-1-b					
Sieve Analysis (percent passing) No. 10 No. 40 No. 200	50 max. 30 max. 15 max.	50 max. 25 max.	51 min. 10 max.	35 max.	35 max.	35 max.	35 max.
Characteristics of fraction passing No. 40 Liquid limit Plasticity index	6 max.		NP	40 max. 10 max.	41 max. 10 max.	40 max. 11 min.	41 max. 11 min.
Usual types of significant constituent materials	Stone fragments, gravel and sand		Fine Sand	Silty or clayey gravel and sand			
General subgrade rating	Excellent to good						

Figure 2.11 AASHTO Soil Classification System.

General Classification	Silt–clay materials (More than 35% of total sample passing No. 200)			
Group Classification	A–4	A–5	A–6	A–7 A–7–5[a] A–7–6[a]
Sieve Analysis (percent passing) No. 10 No. 40 No. 200	36 min.	36 min.	36 min.	36 min.
Characteristics of fraction passing No. 40 Liquid limit Plasticity index	40 max. 10 max.	41 min. 10 max.	40 max. 11 min.	41 min. 11 min.
Usual types of significant constituent materials	Silty soils		Clayey soils	
General subgrade rating	Fair to poor			

[a]For A–7–5, PI < LL – 30
[b]For A–7–6, PI < LL – 30

Figure 2.11   Continued

Figure 2.11 is the work sheet for the AASHTO soil classification system. The data are applied from the left side toward the right. The first group from the right into which the test data fit is the correct classification.

2.7.3 U.S. Comprehensive Soil Classification System

This system was developed by the U.S. Department of Agriculture to organize soils into established groups, identify their best uses, and to enable estimates of crop productivity. The USDA system is very thorough and allows precise identification of almost every soil unit. It is best suited, however, for agricultural use.

The USDA system employs a taxonomic procedure, not unlike that used for biological classification. Six levels of organization are used (see Table 2.5). A brief description of each level is presented below.

Order

Ten soil orders are recognized. The properties used to differentiate between orders are those which tend to give broad climatic groupings of soils. Two exceptions are the Entisols and Histosols, which occur in many different climates. Each order is named with a word of three or four syllables ending in *sol* (for example: Entisol).

Suborder

Each order is divided into suborders that are based mainly on those characteristics that seem to produce classes having the greatest genetic similarities. Suborders narrow the broad climatic range of the orders. Soil properties used to separate suborders are mainly those that reflect either the presence or absence of water logging or soil differences that result from the climate or vegetation. The names of suborders have two syllables. The last syllable indicates the order. An example is *Aquent* (Aqu, meaning water, and ent from Entisol).

Great Group

Each suborder is divided into great groups based on uniformity in the kinds and sequence of major soil horizons and features. The horizons used to make separations are those in which clay, iron, or humus have accumulated; have hardpans that interfere with root growth or water movement (or both); and have thick, dark-colored surface horizons. The features used are the self-mulching properties of clay, soil temperature, major differences in chemical composition, dark-red and dark-brown colors associated with basic rocks and the like. Names of the great groups have three or four syllables and are made by adding a prefix to

Table 2.5 Orders of the U.S. Comprehensive Soil Classification System

Order	Derivation	Formative Element	Description
Alfisols	Meaningless syllable	alf	Gray to brown epipedons; formed mostly in humid region areas under native deciduous forests
Aridisols	L. "aridus" dry	id	Desert or dry soils with an ochric epipedon
Entisols	Meaningless syllable	ent	Recent soils lacking profile development; found under wide variety of climatic conditions
Histosols	Gr. "histos", tissue	ist	Organic soils
Inceptisols	L. "inceptum",	ept	Young soils, more developed than Entisols; Dark (mollic) epidons, includes some of the world's most important agricultural soils
Mollisols	L. "mollis", soft	oll	Dark, (mollic epipedons, includes some of the world's most important agricultural soils
Oxisols	Fr. "oxide", oxide	ox	Oxic subsurface horizon from intense leaching of silica leaving Fe and Al oxides
Spodosols	Gr. "spodos", wood ash	od	Lighter colored (usually albic) horizon above spodic horizon
Ultisols	L. "ultimus", last	ult	Old, moist soils developed under warm to tropical climates, argillic horizons with low base saturation
Vertisols	L. "verto", turn	ert	High content of swelling clays which can develop deep, wide cracks when dry

the name of the suborder. An example is *Haplaquents* (Hapl, meaning simple horizons; aqu for wetness; and ent, from Entisols).

Subgroup

Each great group can be divided into subgroups, one that represents the central (typic) segment of the group and others, such as intergrades (transitional forms to other great groups), and extra grades that have some properties which are representative of the great groups but do not indicate transition to any other kind of soil. The names of subgroups are derived by placing one or more adjectives in front of the name of the great group. An example is *Typic Haplaquents* (a typical Haplaquent).

Family

Soil families are established within a subgroup mainly based on properties important to the growth of plants or on the behavior of soils when used for engineering. Among the properties considered are texture, mineralogy, reaction, soil temperature, permeability, thickness of horizons, and consistency. A family name consists of a series of adjectives preceding the subgroup name. The adjectives are the class names for texture, mineralogy, and so on, used to define family differences. An example is the coarse-loamy, siliceous acid thermic family of Typic Haplaquents.

Series

The series consists of a group of soils that formed in a particular kind of parent material and have genetic horizons that, except for texture of the surface layer, are similar in differentiating characteristics and in their arrangement in the soil profile. Among these characteristics are color, texture, structure, consistence, reaction, and mineral and chemical composition.

2.8   UTILIZATION OF CLASSIFICATION SYSTEMS

The preceding paragraphs present the most common systems of soil classification, and the reader may well ask "what does this all mean?" Any procedure for orderly classification is designed to identify the similarities of a particular soil in such detail that it can be compared to other soil samples. Many tests conducted as part of the classification process provide useful information to the environmental specialist. Several examples of transferable data are listed below.

Particle size distribution provides an indication of the gross surface area of mineral particles which may be available for reaction. The slope of the particle size distribution curve indicates the possible packing density of the soil that is directly related to permeability. *Color* is an indication of the oxidation state of the minerals (bright colors suggest oxidizing conditions, gray or dull colors are typical of reducing environments). Other useful adaptations of descriptive data will be presented in following chapters.

Investigation of each particular site is a necessary precursor to any remedial evaluation or restoration. Accumulation of adequate data for this preliminary phase is a crucial part of the field work, which can be limited by restrictions of time or financial resources. When sources of high quality data (i.e., USDA or civil engineering studies) are available from the local (or similar) area, these can often be used to reinforce data collected specifically from the project site. Often, published data have been subjected to critical review by users, enhancing its reliability. The wise investigator will integrate as much data as possible into the project report. The best reports leave little opportunity for criticism of omission.

## CHAPTER 3

Soil Mineralogy

3.1   INTRODUCTION

Most soils are composed of both organic and inorganic compounds. One half to two thirds of the overall soil volume is comprised of solid matter. Typical soils contain predominantly inorganic compounds that range from 40 to 60% of the soil's total volume (with the exception of organic soils such as peat and muck).

Soil organic matter is derived primarily from the partial decomposition of plant tissue. Mineral grains are solid particles which reflect the developmental history of the soil.

Minerals are solid, naturally occurring chemical compounds which have been formed mostly through inorganic processes. The study of mineralogy includes the identification of minerals, their physical and chemical properties, origins, and classification. Subdivisions include chemical composition, physical properties, and crystallography. All of these disciplines are highly interrelated.

In thermodynamic terms, the minerals present in a soil reflect the current state of total energy in the local soil system. Individual minerals present represent an effort to establish chemical equilibrium with the prevailing local environment. Thus, soils while seemingly stable, are, in reality, in constant transition. Some changes are effected by the processes of weathering. Chemical reactions are facilitated by water (i.e., rainwater saturated with carbon dioxide). Mechanical weathering can occur by thermal stress (freeze and thaw) or by mechanical stress (windblown particles). Other alterations are the result of biological activity (microbial action or plant enzymes). Few places on earth are devoid of some form of continuing soil development.

3.2   CRYSTAL FORMATION

Minerals are the solid form of natural chemicals. All atoms (as ions) have some affinity to combine with other atoms. The concept of valence is a key ingredient in this action. The outer electron orbit of most common elements has eight positions for electrons (except hydrogen, which has only one electron). Chemicals which have fewer than four electrons in the outer orbit are said to be positive (i.e., Ca^{2+}, having a valence of +2), while chemicals that have more than four are considered negative (i.e., Cl^-, which represents seven electrons in the outer orbit). Elements that have four electrons in the outer orbit, such as C and Si, can be either positive or negative; thus, they are capable of building complex compound (and crystal) structures.

When crystals form, the atoms (as ions) present in the system make every attempt to satisfy their electrical balance by combining with the other elements available. As the atoms combine into three-dimensional solid form (crystals), they assemble into an orderly form. The form of this ordered atomic arrangement is determined partly by the charge of the atoms, but even more so by the size of the atoms involved. Figure 3.1 shows the relative size of some of the most common mineral-forming ions.

Any time a mineral crystal is allowed to form without space restrictions it will develop individual crystals with well-formed crystal faces which reflect its internal structure. Most of the time, however, crystal growth is interrupted by competition for space. This results in an intergrown mass of crystals, none of which exhibit visual crystal form. Soil grains are almost always weathered to a degree that their crystal structure is not visibly apparent.

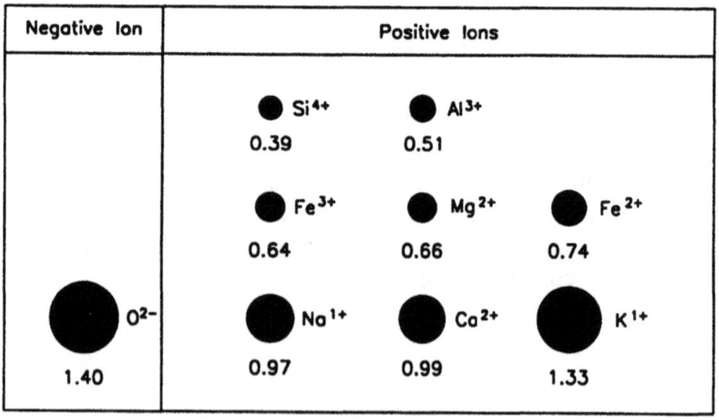

Negative Ion	Positive Ions		
	Si^{4+} 0.39	Al^{3+} 0.51	
	Fe^{3+} 0.64	Mg^{2+} 0.66	Fe^{2+} 0.74
O^{2-} 1.40	Na^{1+} 0.97	Ca^{2+} 0.99	K^{1+} 1.33

Figure 3.1   Relative sizes of common mineral-forming ions. Diameter shown is in angstroms (one angstrom = 1 x 10^{-8} cm).

In some crystals, weakness planes occur where the bonding strength between adjacent layers of atoms is relatively weak. These planes also define the external form of the crystal when it is allowed to grow in an unrestricted space. Cleavage (fracture) is more likely to occur along these weak planes, between the rows of atoms. When a diamond is "cut," it is cleaved along selected planes to enhance light refraction. Weathering (degradation) tends to start on cleavage planes and corners of the crystals where exposure is great and strength is low.

Crystal structure (internal and external) is determined by the alignment of atoms in the crystal. The six basic crystal classes (cubic, hexagonal, rhombohedral, monoclinic, orthorhombic monoclinic, and triclinic) all are the result of the packing organization of ions in the crystal. The density (unit weight per volume) of each crystal reflects the atomic weight and packing (organization) of contained atoms.

During crystal formation, some basic forms of crystal structure are very common. The silica tetrahedron is formed when silicon and oxygen (in an abundance of oxygen) combine with other elements or compounds. Si^{4+} is able to join with four oxygen atoms (each 2-) because each oxygen uses one of its two electron positions (or valence) to attach to the Si, while the other electron position is available to attach with an adjacent atom. In three-dimensional form, the shape is a tetrahedron shown in Figure 3.2.

Another common structure is the combination of Al^{3+} with O^{2-} in the form of an octagonal structure (Figure 3.2). Aluminum octahedra are *almost* (but not quite) the same size as two adjacent silica tetrahedron, but the octahedron has a charge deficiency. The setting becomes complex when ions of differing size and charge have substituted for the Al or Si ion. In many situations, the substitution results in an electrical imbalance which invites other ions to join the structure in an effort to achieve electrical neutrality.

The complexity of crystal structure is shown on Figure 3.3, which is a schematic diagram of the clay mineral kaolinite. Alternating layers of tetrahedra and

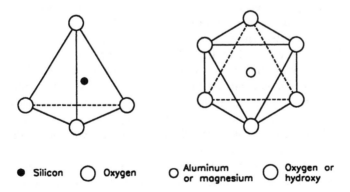

● Silicon   ○ Oxygen   ○ Aluminum or magnesium   ○ Oxygen or hydroxy

Figure 3.2 Basic forms of silicate structure. (a) Tetrahedron, (b) Octahedron. (Adapted from Brady, 1990. Used with permission.)

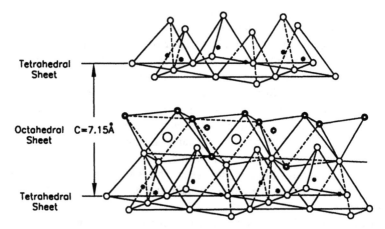

Tetrahedral
Sheet

Octahedral C=7.15Å
Sheet

Tetrahedral
Sheet

Figure 3.3 Diagrammatic representation of a kaolinite crystal structure. Symbols
represent: large heavy circle, oxygen atoms; small circle, silicon atoms; double circle, hydroxyl ions; large light circle, aluminum atoms octrahedrally
coordinated.

octahedra form the basic structure. The relatively weak bonding between the layers results in planes of weakness. Thus, clay minerals form pancake-shaped, flat
crystals. The small size of clay crystals results from a slight difference in the
dimensions of octahedrons and tetrahedrons. After several hundred of these structures are lined up next to each other, the minor size difference accumulates to
restrict the matchup and crystal growth stops.

The almost infinite number of possible ionic substitutions is one reason for
the multitude of minerals identified in the world. Review of any standard mineralogy text will confirm this complexity. Fortunately, only a few common minerals are found in significant quantities in most soils. Figure 3.4 presents examples of the crystal structures of some common silicate minerals.

As an example of mineral formation, consider the salts resulting from evaporation of sea water. The first crystals formed are predominantly sodium chloride (Halite). The crystal form of Halite is cubic because each chloride atom
shares its electron orbits with six adjacent sodium atoms (see Figure 3.5). As
seawater evaporation continues to the point where most of the sodium is consumed, other salts begin to precipitate, including: potassium chloride (KCl), calcium chloride ($CaCl_2$), calcium sulfate (gypsum, $CaSO_4$), as well as minor salts.
Only in the latter stages of dryness do the bromine minerals precipitate, because
bromine is a very small component of seawater.

Evaporation of seawater was used as an example because it is a commonly
observable phenomenon. In other mineral formations, such as those resulting from
high temperature/high pressure silicate melts deep in the earth, the processes of
mineral formation become exceedingly complex. Not only does the abundance

Mineral		Idealized Formula	Cleavage	Silicate Structure
Olivine		$(Mg,Fe)_2SiO_4$	None	Single tetrahedron
Pyroxene		$(Mg,Fe)SiO_3$	Two planes at right angles	Chains 
Amphibole		$(Ca_2Mg_5)Si_8O_{22}(OH)_2$	Two planes at 60° and 90°	Double chains 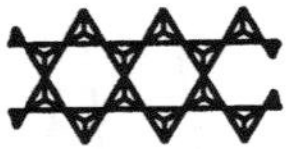
Micas	Muscovite	$KAl_3Si_3O_{10}(OH)_2$	One plane	Sheets 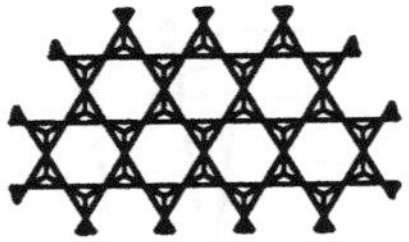
	Biotite	$K(Mg,Fe)_3Si_3O_{10}(OH)_2$		
Feld-spars	Orthoclase	$KAlSi_3O_8$	Two planes at 90°	Three-dimensional networks 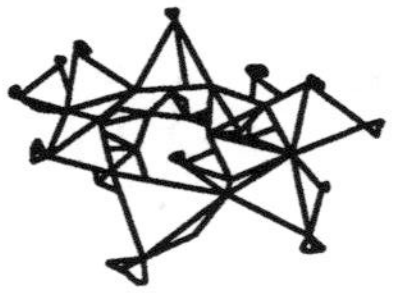
	Plagioclase	$(Ca,Na)AlSi_3O_8$		
Quartz		SiO_2	None	

Figure 3.4 Common silicate minerals. Note that the complexity of the silicate structure increases down the chart.

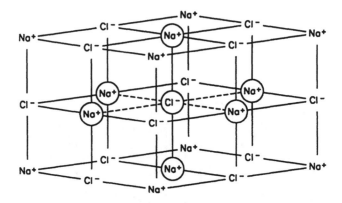

Figure 3.5 Schematic illustrating the arrangement of sodium and chloride ions in table salt (all Na have a "+" and all Cl have a "–").

of an element become important, but also its chemical activity, ionic size, and thermal stability. As the temperature and pressures change, so do the mineral forms. Later, when these high energy environment minerals are exposed at the surface of the earth, they are vulnerable to alteration as they adapt to the new environment.

3.3   CHEMICAL MINERAL RELATIONSHIPS

Minerals in soil can also be subdivided into two general mineral groups reflecting their process of formation: primary minerals and secondary minerals. Primary minerals are typically inherited from parent materials. In order of decreasing abundance, primary minerals include quartz, feldspar, and mica, and subordinate amounts of heavy minerals including amphibole, pyroxene, olivine, epidote, tourmaline, zircon, and rutile. Such minerals are comprised of fundamental silicate units. Quartz is abundant in sands and silts. In addition to quartz, feldspar varieties are also abundant where leaching has not been extensive. Mica is a source of potassium in most temperate-zone soils. Amphiboles are easily weathered to clay minerals and oxides. Pyroxene and olivine are both easily weathered. These minerals and their respective chemical formulas are presented in Table 3.1

Secondary minerals typically result from the weathering and transformation of primary silicates. They are typically clay size and characterized by a relative lack of order in atomic structure. Minerals such as kaolinite, smectite,

Table 3.1   Common Minerals Found in Soils

Mineral Name	Chemical Formula
Primary Minerals	
Quartz	SiO_2
Feldspar	$(Na,K)AlO_2CaAl2O_4[SiO_2]_2$
Mica	$K_2Al_2O_5[Si_2O_5]_3Al_4(OH)_4$ $K_2Al_2O_5[Si_2O_5]_3(Mg,Fe)_6(OH)_4$
Amphibole	$(Ca,Na,K)_{20}3(Mg,Fe,Al)_5(OH)_2$ $[Si,AlO_{11}]_2$
Pyroxene	$(Ca,Fe,Mg,Ti,Al)(Si,Al)O_3$
Olivine	$(Mg,Fe)_2SiO_4$
Epidote	$Ca_2(Al,Fe)_3(OH)Si_3O_{12}$
Tourmaline	$NaMg_3Al_6B_3Si_6O_{27}(OH,F)_4$
Zircon	$ZrSiO_4$
Rutile	TiO_2
Secondary Minerals	
Kaolinite	$Si_4Al_4O_{10}(OH)_8$
Smectite[a]	$M_x(Si,Al)_8(Al,Fe,Mg)_4O_{20}(OH)_4$
Chlorite	$M_x(Si,Al)_8(Al,Fe,Mg)_4O_{20}(OH)_4$
Allophone	$Si_3Al_4O_{12}\, nH_2O$
Imogolite	$Si_2Al_4O_{10}\, 5H_2O$
Gibbsite	$Al(OH)_3$
Goethite	$FeO(OH)$
Hematite	Fe_2O_3
Ferrihydrite	$Fe_{10}O_{15}\, 9H_2O$
Birnesite	$(Na,Ca)Mn_7O_{14}\, 8H_2O$
Calcite	$CaCO_3$
Gypsum	$CaSO_4\, 2H_2O$

[a]M = Interlayer Cation.

vermiculite, and chlorite are abundant in clay soils as products of weathering. These same minerals are also sources of exchangeable cations in soils. Allophone and imogolite (less common clay minerals) are abundant in soils derived from volcanic ash deposits. Gibbsite is abundant in leached soils. Ferrihydrite is abundant in organic horizons.

The most abundant Fe and Mn oxides are goethite and birnesite, respectively, reflecting their persistence in the soil environment relative to the secondary silicates. Elements such as Al, Fe, or Mn are less readily leached than Si, unless appreciable amounts of soluble organic matter (as acids) are present which render the metals more mobile. Calcite is the most abundant carbonate. In warm, arid regions, hematite and gypsum are most abundant, respectively. These minerals and their respective chemical formulas are also presented in Table 3.1.

Mineralogical properties are often related to grain size. Sands are generally composed, primarily, of hard and stable minerals and may consist mainly of quartz (plus a small percentage of other minerals). Factors such as source areas, methods of transport, distance transported, and climate all play a key role in the resulting mineralogy of soils. Soils derived from oceanic islands may produce sands consisting mainly of calcite, whereas ash (silica) may be the primary constituent derived from volcanic source areas. Glacial sands, on the other hand, may contain minerals that are prone to weathering and exposure, due to their relatively young age. Silts can contain both inorganic (mineral) and/or organic (decaying plant matter) constituents. Clay particles (less than .002 mm in size) can also be a main constituent of soils, and are mainly comprised of clay minerals. Typical clay minerals present determine the basic characteristics (both physical and chemical) of fine grained soils.

Organic matter, or humic substances, is also an important constituent in soils. Humic substances are dark colored microbially transformed organic materials which persist in soils throughout profile development. The two most common humic substances are humic and fulvic acid. These acids are enriched in carbon but depleted in hydrogen and nitrogen; the depletion of nitrogen reflects the absence of constituents that are susceptible to net mineralization via biodegradation. The average chemical formulas for humic and fulvic acids are presented in Table 3.2.

3.3.1   Primary Silicates

Primary silicates in soils are derived from the physical disintegration of parent rock material. Important primary silicate minerals and their respective mineral groups are summarized in Table 3.3.

Table 3.2   Chemical Formulas for Soil Organic Matter

Material	Chemical Formula
Humic Acid	$C_{187}H_{186}O_{89}N_9S$
Fulvic Acid	$C_{135}H_{182}O_{95}N_5S_2$

Table 3.3 Primary Silicate Minerals and Mineral Groups

Mineral Group	Mineral Name	Chemical Formula
Silicate	Quartz	SiO_2
Olivine	Fayalite	Fe_2SiO_4
	Forsterite	Mg_2SiO_4
	Chrysolite	$Mg_{1.8}Fe_{0.2}SiO_4$
Pyroxine	Diopside	$CaMgSi_2O_6$
	Enotatite	$MgSiO_3$
	Ortho Ferrosilite	$FeSiO_3$
Amphibole	Actinolite	$CaMg_4FeSi_8O_{22}(OH)_2$
	Hornblende	$NaCa_2Mg_5Fe_2AlSi_7O_{22}(OH)$
	Tremolite	$Ca_2Mg_5Si_8O_{22}(OH)_2$
Mica	Biotite	$K_2[Si_6Al_2]Mg_4Fe_2O_{20}(OH)$
	Muscovite	$K_2[Si_6Al_2]Al_4O_{20}(OH)_4$
	Philogopite	$K_2[Si_6Al_2]Mg_6O_{20}(OH)_4$
Feldspar	Albite	$NaAlSi_3O_8$
	Anorthite	$CaAl_2Si_3O_8$
	Orthoclase	$KAlSi_3O_8$

Minerals of the olivine group are comprised of individual silica tetrahedron clusters which make up the members: fayalite, forsterite, and chrysolite. Solid solution occurs between fayalite and forsterite to produce a series of mixtures with specific names (i.e., the mineral chrysolite, for example, contains 10 to 30 molecular percent fayalite). Weathering of olivine minerals is relatively rapid and tends to start along cracks and along crystal surfaces (especially corners). Such weathering can form altered ring-shaped patterns containing solid phase oxidized iron and smectite.

Pyroxenes and amphiboles are comprised of single and double chains of silica tetrahedra. Weathering of these silicates results in the formation of smectite, similar to that of olivine.

Micas are comprised of sheets of silica tetrahedra, fused to each plane side of a sheet of metal cation octahedra. Muscovite and biotite are the most common soil micas. Muscovite structure is dioctahedral (two octahedrons together), whereas the cation (inside the structure) is trivalent. Only two of the three possible cationic sites in the octahedral sheet are filled to achieve a charge balance. Should the metal cation be bivalent, all three possible sites are filled and the sheet is termed dratahedral, as in the case of biotite. Muscovite typically weathers to vermiculite, which continues to weather to smectite. Biotite also weathers to vermiculite and geothite.

Feldspars are continuous, three-dimensional framework of silica tetrahedra with shared corners. Some of the tetrahedra contain Al instead of Si. Feldspar initially decomposes to produce allophone and smectite, and may eventually weather to the clay minerals kaolinite and gibbsite.

3.3.2   Oxides and Hydroxides

Oxide minerals are comprised of octahedral structures with a preponderance of iron, aluminum, or manganese rather than silica. Goethite is the most thermodynamically stable of the iron oxides, regardless of the climatic region. Gibbsite is the most important aluminum mineral, whereas birnesite is the most common manganese mineral. Oxides and hydroxides represent the end product, and form either from the weathering of primary silicates or from the hydrolysis and desilication of certain clay minerals (i.e., smectite and kaolinite). A summary of the most important oxides and hydroxide minerals is presented in Table 3.3.

3.3.3   Carbonates

Carbonate minerals found in soils include calcite ($CaCO_3$), dolomite [$CaMg(CO_3)_2$], nahcolite ($NaHCO_3$), trona [$Na_3HC(CO_3)_2\,2H_2O$], and soda ($Na_2CO_2\,10H_2O$). Calcite can be either a primary or secondary mineral in soils. Secondary calcite forms as a precipitate from solutions enriched in soluble Mg (which forces the Ca out of solution). Secondary calcite that co-precipitates with $MgCO_3$ forms a magnesium calcite which accounts for nearly all the secondary Mg carbonate in soils. Secondary calcite can also occur as coatings on other minerals. Pedozenic (found in soil) calcite occurs through the weathering of Ca-bearing primary silicates (i.e., pyroxenes, amphiboles and feldspars), and primary calcite.

3.3.4   Sulfates

Sulfate materials occur as a by-product of weathering under arid conditions. The principal sulfate minerals include gypsum ($CaSO_4\,2H_2O$), anhydrite ($CaSO_4$), epsomite ($MgSO_4\,7H_2O$), mirabilite ($Na_2SO_4\,10H_2O$), and thenardite (Na_2SO_4). In a manner similar to calcite, gypsum can dissolve and re-precipitate in a soil profile via leaching by rainwater, irrigation water, or injected water. Some varieties of sulfates similar to the carbonates form within the upper portion of the soil profile as it dries.

Other sulfate minerals can be produced by precipitation in acid soils where the sulfate is derived via sulfur oxidation or by the addition of gypsum, which reacts with abundant Fe and Al in the soil solution. Such mineral precipitates include jarosite [$KFe_3(OH)_6(SO_4)_2$], alunite [$KAl_3(OH)_6(SO_4)_2$], basaluminite [$Al_4((OH)_{10}SO_4\,5H_2O)$], or jurbanite ($AlOHSO_4\,5H_2O$). These minerals in turn can dissolve incongruently, forming ferrihydrite or gibbsite.

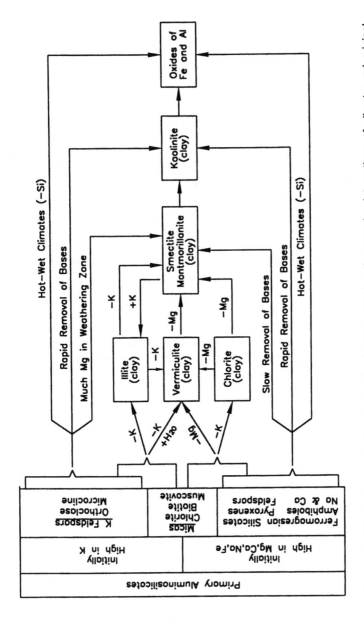

Figure 3.6 Weathering products of primary minerals, showing the complex interactions of climate and original minerals. (Adapted from Brady, N.C., *The Nature and Properties of Soil*, 10th ed., Macmillan Publishing Co., New York, 1990. Used with permission.)

3.4   MINERAL WEATHERING

All materials are subject to weathering. A common example is that of concrete. Freshly placed concrete appears solid and permanent. However, after only a few years, it is often cracked, with a very rough appearance. The same natural forces that degrade concrete also reduce rocks to soil and continue to alter the minerals in the soil.

Alteration of minerals is caused by chemical or mechanical weathering, and often a combination of the two. Mechanical weathering can be a product of abrasion (by water or wind forces) or frost wedging in cracks in the soil grains (freeze and thaw). In many cases, mechanical breakdown of mineral grains is accentuated after the grains have been weakened by chemical action.

Chemical weathering of soil grains involves complex processes that alter the internal structure of minerals by removing and/or adding elements. During the transformation, the substances are in balance (equilibrium) with the present environment. Soil, however, is exposed to rainfall, biological activity, aeration, and heating/cooling, and thus, is seldom in perfect equilibrium.

Water is the most active agent in chemical weathering. While pure water is almost nonreactive, only a small quantity of dissolved material is all that is necessary to activate it. The major processes by which water alters minerals are: solution, oxidation, and hydrolysis. Figure 3.6 describes the results of chemical weathering of primary minerals. Each of these processes is discussed in greater detail in Chapter 6, Soil Chemistry.

Different varieties of minerals are less or more stable, depending on their crystal structure, chemical composition, and surface area. Some minerals appear to be stable during human lifetimes, while others are observably altered in a relatively short period of time.

## CHAPTER 4

Soil Mechanics

4.1 INTRODUCTION

Soil is a term which develops a mental image of a body of material, usually in the subsurface, which contains solid matter, water, gases, and possibly other components such as living tissue, or man-made inclusions. Chemical and physical reactions within and between these individual components are often very complex. The response of a "bulk" volume of soil to both internal and external mechanical forces is also an important determining factor of how a soil exists and functions in the environment.

A crude analogy for soil mechanics is that of a house. The house may be constructed of wood, brick, concrete, or other materials, but it is not a solid structure. Included in the spatial volume occupied by the house are solid walls, floor, roof, plumbing, electrical fixtures, and a large volume of void space. With all of these components, the house is still recognized as a house that responds as a unit to a myriad of external and internal forces.

This chapter explores some of the physical attributes of a bulk soil and how the soil responds to mechanical forces. Concepts of density, volume, stress-strain, and applications of these factors (as slope stability and compaction) are presented, along with example calculations.

4.2 SPACE AND VOLUME RELATIONSHIPS

All natural soil consists of at least three primary components, solid (mineral), void (not occupied by mineral), and water. The relationship between these components is shown in Figure 4.1. For convenience and clarity, each major component has been reduced to a concentrated commodity within a unit volume. Proportions of these components vary dramatically between and within various soil types. Water (not chemically attached) is in reality a "void filler," and

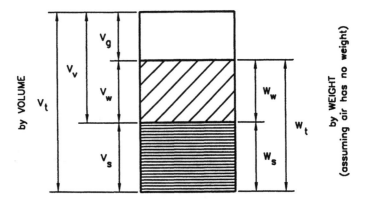

Figure 4.1   Weight/volume relationship of soil.

the relationship between it and void are dependent on how much moisture is available.

Within a given soil, the proportions of the three major components may be mechanically adjusted by reorientation of the mineral grains by compaction or tilling. Occasionally, it is prudent to blend soil types to alter the proportions for a specific purpose, such as increasing (or decreasing) the percentage of void space.

Porosity (η) is the term used to define the ratio of the volume of void spaces to the total volume. Usually, porosity is expressed as a percentage.

$$V_v/V_t = \eta$$

where:

$\quad \eta$ = Porosity
$\quad V_v$ = Volume of voids
$\quad V_t$ = Total volume of soil

Void ratio (e) is the ratio of volume of voids to volume of solids.

$$V_v/V_s = e$$

where:

$\quad$ e = void ratio
$\quad V_v$ = Volume of voids
$\quad V_s$ = Volume of solids

If a unit volume of space is filled with uniformly sized spheres, the porosity depends on the packing orientation of those spheres. In the densest state = 26%,

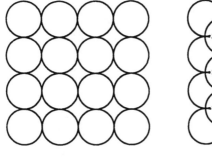 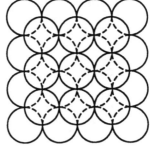

(a) Very loose, (Porosity=.48)   (b) Very dense, (Porosity=.26)

Figure 4.2 Packing density of soil particles.

while at the loosest state, $\eta = 47\%$ (see Figure 4.2). The porosity of natural sand varies between about 25 to 50%, depending on the shape of the grains, distribution of grain sizes, and the conditions of sedimentation. The effect of varying grain sizes can easily be visualized by imagining blending small lead shot (1 mm) with BB shot (5 mm) and marbles (12 mm). BBs will fit readily between the marbles, and lead shot can be packaged between the BBs. When these spheres are arranged into the densest packing, the void space is significantly reduced.

The porosity of natural clays and other soils which contain a large percentage of flat particles can be as high as 60% or greater. This phenomenon is caused by the large surface area to volume ratio of the clay particles. Water attached to the surfaces of these particles often limits the area of particle to particle contact, thus keeping the particles apart. It is not uncommon to find uncompacted clay soils with void ratios significantly greater than one.

Moisture content is defined, for civil engineering purposes, as the ratio of the weight of water to the dry weight of mineral aggregate. It is expressed as a percentage.

$$\frac{W_s}{W_w} = \%\text{Moisture}$$

Where:
W_s = Dry weight soil
W_w = Weight of Water

In soils above the water table, some of the voids are filled with air. Partial saturation can be calculated by the equation:

$$V_w/V_v = \% \text{ Saturation}$$

Table 4.1   Specific Gravity of Soil Minerals

Mineral	Specific Gravity
Quartz	2.65
Kaolinite	2.6
Dolomite	2.87
Gypsum	2.32
Calcite	2.72
Illite	2.8
Montmorillonite	2.65–2.80
Potassium feldspar	2.57
Sodium and calcium feldspar	2.62–2.76
Biotite	2.8–3.2
Muscovite	2.76–3.1
Horneblende	3.0–3.5
Limonite	3.6–4.0
Olivir.e	3.27–3.37

where:

V_w = Volume of water

V_v = Volume of voids

The unit weight of a soil is the total weight of aggregate plus water contained within a unit volume of soil. It is usually expressed as pounds per cubic foot or kilograms per cubic meter. Unit weight is dependent upon: the weight of aggregate, weight of water, porosity, and degree of saturation.

Evaluation of weight/volume relationships in soil requires that the specific gravity of the solid aggregate components be known, so that their volume can be calculated. Table 4.1 presents the densities of some of the most common minerals found in soil.

Examples 4.1 and 4.2 present procedures for calculating the weight /volume relationships. The purpose of these calculations is to determine: Is this soil saturated? What is the void ratio? Is it reasonable to consider further compaction to increase density?

4.3   SOIL REACTIONS

4.31   Effective Stress

In a large mass of soil, the pressure within the soil increases as the depth increases. A soil that has a unit weight of 100 pounds per cubic feet exerts a pressure of 100 psf at one foot depth and 300 psf at three feet, etc. Throughout the unit, the structure of the soil grains reorient themselves to support the cumulative added weight. Due to the effects of elasticity, a sample retrieved from a specified depth may not be truly representative once it is delivered to the surface, where it is no longer subjected to its confining load.

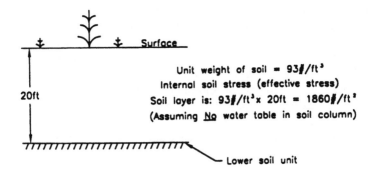

Unit weight of soil = 93#/ft³
Internal soil stress (effective stress)
Soil layer is: 93#/ft³ x 20ft = 1860#/ft²
(Assuming No water table in soil column)

Figure 4.3a   Effective stress on lower soil unit, without water table.

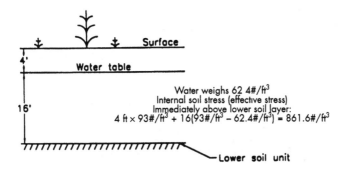

Water weighs 62 4#/ft³
Internal soil stress (effective stress)
Immediately above lower soil layer:
4 ft × 93#/ft³ + 16(93#/ft³ − 62.4#/ft³) = 861.6#/ft³

The bouyancy of the water greatly
reduces the effective stress (internal soil stress).
Note how lowering the water table greatly increases
the soil stress on objects located at depth.

Figure 4.3b   Effective stress on lower soil unit, with water table.

If a load of heavy material (i.e., a gravel pile) is placed on an unsaturated soil, the soil tends to be compressed in direct response to its load. The internal organization of the aggregate particles reorganize to support the added load and the internal shear strength increases accordingly. The added load to the soil is termed *effective stress* (see Figure 4.3).

If the same soil sample is placed in a container and water is added as the load (saturates the soil, but cannot escape), the soil is not comparably compressed; instead, the water adds buoyancy and the internal shear stress is not increased. This type of loading is considered neutral stress. Under these circumstances, the effective pressure of one soil layer of one soil unit resting upon another is reduced by the unit weight of water. This relationship is described in Figure 4.3.

The value of submerged unit weight and effective stress to environmental specialists is often experienced during excavation or drilling activities, particularly in sand soils (with no cohesive strength). If a sand layer below the water table is continually dewatered during excavation, the combination of inflowing groundwater and reduced effective stress (weight of soil on the bottom) may result in "heaving" or "boiling," as shown in Figure 4.4. This situation can be controlled by (1) maintaining the water level near or at the level of the surrounding groundwater or (2) lowering the local water table to the level of the bottom of the excavation, thus reducing the effective stress adjacent to the hole.

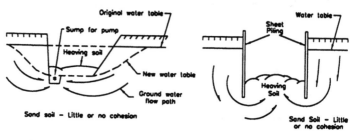

Soil Heave in an Excavation   Soil Heave in "Shored" Excavation

In each example above, the weight of the outside soil and movement of groundwater have exceeded the shear strength of the soil at the base of the excavation.

Figure 4.4a   Soil heaving into an excavation.

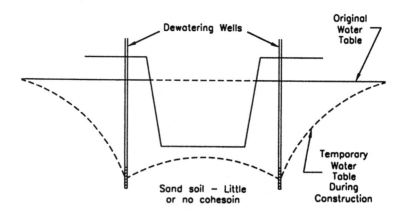

When the water table is lowered below the excavation, the sand may be excavated without heaving.

Figure 4.4b   Dewatering to reduce soil heaving.

Similar conditions are observed when auger drilling in sands below the water table. When the center rod or core barrel is withdrawn from the center of the auger (or other drill tools), it often reduces the bottom hole stress on the soil, and the sand "runs up the hole." This occurrence can be reduced or eliminated by maintaining the fluid level in the bore of the auger higher than the ground-water table (to maintain a "positive stress").

4.3.2   Compressibility of Soil

When an added weight (load) is placed above a soil layer, some settlement can be expected. The vertical load may be caused by construction of a building, piling soil to construct a dike, material stockpile, or construction of an above-ground landfill.

The soils responded to the added weight by reorienting the aggregate parti-cles into a structure which has sufficient sheer strength to sustain the added load. This compressibility is most significant in clay soils because of the inherent high porosity. In noncohesive (sand and gravel) soils, the effect is less important be-cause the particles tend to be more spherical and are often deposited in a rela-tively dense state; thus, the added load is borne by direct grain-to-grain contact. Mixtures of soils respond to loading depending on their natural void ratio, mois-ture content, and state of previous loadings.

When loading occurs, soils below the surface layer seldom respond imme-diately. Over a period of time the soil grains are reorganized and void ratio re-duced. The settlement results in the extrusion of water or displacement of air. The rate of settlement is highly dependent on the soil type, depth of the com-pressible layer, and the ease with which contained fluids can be displaced. Con-fined clay layers at significant depth may respond very slowly because contained water may not have an escape route.

The mechanics of settlement are quite complex and require detailed site data, extensive soils laboratory analysis, and careful calculation to assure accurate pre-dictions. For detailed discussions, the reader is referred to the standard text ref-erences at the end of the book.

4.3.3   Stress and Strain in Soils

The response of a soil to *stress* (pressure) is called *strain* (displacement). The relationships between stress and strain determine the settlement of foundations and the change in earth pressure due to small movements of retaining walls and bracing in earth cuts. Mechanical strength of soil results from two factors: the resistance of soil particles to displacement around or past each other (resulting from grain to grain contact), and cohesion (sticking together) caused by the sur-face action of individual minerals and water, usually related to clay content.

If settlement of a foundation or structure is due solely to the compression of a soft clay layer at a shallow depth, little horizontal displacement occurs and the

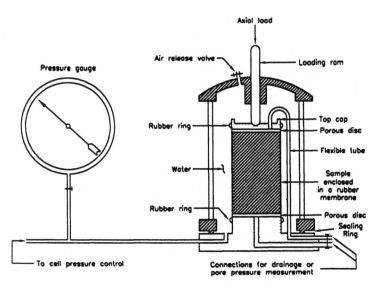

Figure 4.5   Triaxial testing apparatus.

vertical extent can be calculated with reasonable accuracy. Under more complex conditions, the application of a load causes horizontal yield in every direction.

The most common procedure to determine the stress-strain relationship of a soil is the use of a triaxial testing apparatus. This device is diagrammatically shown in Figure 4.5. A prepared soil core is placed in a plastic, watertight membrane (generally round), and a predetermined pressure is established uniformly around the cell. The piston at the top is used to apply a vertical axial load to the top of the sample. As the vertical load is increased, the horizontal yield is recorded. The porous plate at the base of the cell provides an escape route for water contained within the sample.

This type of testing apparatus can be adjusted to approximate field conditions. For example, if the field situation has a sand layer which may allow water drainage, the valve at the base of the apparatus can be opened, allowing a "drained test." Conversely, the test can also be operated as an "undrained test." Horizontal pressure can be set to equal those expected at the depth of concern.

With modification, the test apparatus can be assembled to allow measurement of strain in all four horizontal directions for determination of specific conditions.

Soils, like most solid materials, fail either in compression or tension. Since soils are composed of small (relatively) grains which have minimum cohesion at best, tension strength is minimal. Tension causes cracks to open which are not beneficial. Compression of a sample to failure causes shear planes to develop, which results in lateral spreading of the material. A typical stress-strain curve for an unconfined soil sample is presented in Figure 4.6. Note that the peak stress point is followed by a rapid decline as the strain increases.

a) Relationship between stress and strain on an unconfined clay soil sample. Notice the rapid decline in stress after "failure".

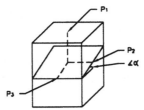

b) Internal stress on a soil sample — Normal (P₁) and shearing stress (P₂ & P₃) on a plane inclined at an angle.

Figure 4.6 Typical stress-strain response for an unconfined soil sample. (Redrawn from Terzaghi and Peck, 1968).

The principles of mechanics demonstrate that the normal and shearing stresses within a sample are related by the equation:

$$p = 1/2 \ (p1 + p2) + 1/2 \ (p3 - p2) \cos 2a$$

This equation can be graphed on a rectangular system in which the horizontal axis represents the normal forces and the vertical axis the shear response (see Figure 4.7). Since triaxial testing almost always uses an equal confining pressure around the horizontal sides of the specimen, $p2 = p3$, only one circle need

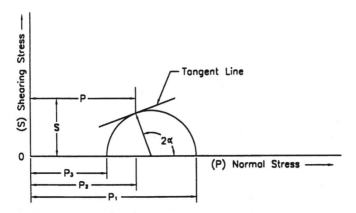

Figure 4.7 Theoretical plot of Mohr's circle of stress plotted from triaxial testing.

be constructed for each test set. Every point on the circle of stress represents the normal and shearing stress on a particular plane inclined at an angle "a" to the direction of the plane of the major principal stress.

If the principal stresses p1 and p3 are derived from the sample at the instant of failure, then at least one point on the circle of stress must represent a combination of normal and shearing stresses that lead to failure on some plane through the specimen. If a series of tests are conducted at varying stresses and plotted on the same diagram (Figure 4.8), a line drawn tangentially to the envelope of circles will define the rupture line for the specific soil. This line also approximates the shearing angle "Φ".

Shear strength of soil results from a combination of cohesion and internal shearing resistance. Coulomb, an early soil scientist, approximated this relationship as:

$$s = c + p \tan \Phi$$

where
   s = shear strength (force per unit area)
   c = cohesive strength (force per unit area)
   p = normal stress
   Φ = slope of the rupture envelope line

Figure 4.8 describes the diagram of Mohr's diagrams for two soils, (a) a low cohesion sand and (b) a cohesive clay which has most of its total shear strength due to cohesion.

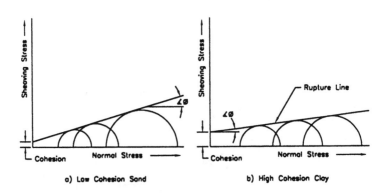

a) Low Cohesion Sand   b) High Cohesion Cloy

Figure 4.8   Plots of actual triaxial tests on two soils. Each test was made at three radial confining pressures ($P_2 = P_3$). A pure sand would exhibit no cohesion, and a pure clay would have a 0 φ angle.

Table 4.2 Drained Angle of Friction (Expressed in
Degrees)

Soil Type	Density	
	Loose	Dense
Sand, rounded grains	27.5	34
Sand, angular grains, well graded	33	45
Sandy gravel	35	50
Sandy silt	27–33	30–34
Inorganic silt	27–30	30–35

Most of the shear strength of sands and organic silts (cohesionless soil) is caused by interlocking of the grains. The values of Φ are not significantly different whether the soil is wet or dry. At normal pressures such as those encountered under small buildings and other light structures, Φ is not significantly affected by differences of loads. This angle closely approximates the load distribution under the surface of the soil. Also, this angle is very close to the "angle of repose" for a stable slope.

Table 4.2 presents typical values for values of Φ for natural soils in the loose and dense state. The ranges presented in this table are only approximate, and should be used with caution.

Pure cohesive soils (clay and plastic silt) often derive most of their shear strength from cohesion between soil grains, and have such a low permeability that water drainage is negligible during the application of stress. Under these conditions, the soil may be considered to have a $\Phi = 0$, and the shear strength is equal to the cohesive strength.

4.4   EFFECTS OF VIBRATION ON SOILS

Vibrations of a soil mass, resulting from operation of machinery, pile driving, traffic, blasting earthquake, or other sources, can cause compaction by reorientation of soil grains. This consolidation results in vertical settlement.

When vibrations are applied to granular soils they often cause water to be displaced from between the grains. If soil is relatively loose, the vibration can result in a "liquefaction" of the soil. This temporary situation is alleviated when the water has been removed or the vibration halted.

4.5   COMPACTION OF SOILS

Discussion in the preceding section focused on the shear strength and stability of soils in their natural state. For earth construction, the stability of soil used as a dike, pond liner, highway base, or foundation depends largely on how the soil was placed. Proper placement procedures will increase shear strength, reduce permeability, or decrease settlement.

The goal of compaction is to increase the soil density (reduce void ratio) and thus increase the shear strength. To accomplish this goal, the soil must be manipulated to reorient the soil grains into a more compact state. Efficient compaction is obtained when the water content is within a limited range which is sufficient to lubricate the particle movement.

Compaction methods for cohesionless soil (sand and gravel) include (in order of effectiveness): vibration, watering, and rolling. A combination of these procedures, such as vibrating roller, is particularly effective.

Vibrations can be produced by such simple means as raising and dropping a weight, or tamping with hand tools. Over large areas, however, the most effective means is by use of equipment vibrating at a frequency close to the resonant frequency for the soil and the vibrator.

The thickness of soil placement layer ("lift") for most effective compaction is usually 6–10 inches (15–25 centimeters). This depth allows the most energy to be transmitted throughout the lift. When greater thicknesses are used, the energy is dispersed and the effort required to accomplish maximum density is *greatly* increased.

As the silt and clay content of soils increases, the compacting effect of vibrations decreases significantly. The bonding between particles caused by cohesion interferes with their reorientation.

Compaction of cohesive soils is best accomplished by "kneading" the soil, similar to blending bread dough. Equipment used for this includes pneumatic tired rollers and sheepsfoot rollers. Pneumatic rollers have a heavily loaded cart supported by a number of inflated rubber tires set apart at approximately 1/2 tire width. As this cart is pulled across the fill area, the soil is worked into a denser state.

For soils which are nearly plastic, a "sheepsfoot" roller is often the best tool. A sheepsfoot is a large diameter 4 foot to 8 foot diameter steel roller with prismatic attachments (feet) attached, one for approximately every 100 square inches. Each of these feet extends at least 6 to 9 inches from the roller. Often the roller is filled with water to increase its weight. As the roller is pulled across the soil, the feet penetrate the soil and cause compaction. While the surface layer appears to be highly pockmarked, the soil below the foot depth is compacted. The maximum compaction of the soil is indicated when the roller is no longer able to penetrate the soil and the "feet" walk on top of the soil surface.

The effectiveness of the compactive effort is determined by testing the in-place dry density of the soil (weight of solids per unit volume). Comparison of the dry density of field-compacted soil to a standard prepared in a laboratory allows a percent compaction comparison.

The laboratory standard is prepared by administering a measured compactive effort to a specified sample volume at different moisture contents. After completion of several laboratory tests, a curve is prepared which presents the dry density versus moisture content (see Figure 4.9). For conditions of the test, the

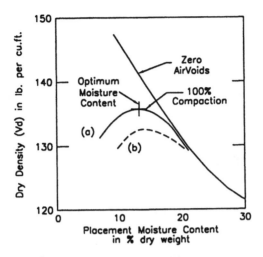

Figure 4.9   Moisture-density (compaction) curve: (a) curve produced by specific compactive effort; (b) curve resulting from less effort.

dry density of the peak of the curve is termed the "maximum dry density," which represents the 100% test value. The moisture content at the peak point of the curve is called the optimum moisture content. It represents the moisture content of the soil which can be most easily worked to reach maximum compaction.

The standard laboratory tests for compaction are the "Test Method Laboratory Compaction Characterization of Soil Using Standard Effort" (Proctor Test— ASTM-D-698) and the "Test Method for Laboratory Compaction of Soil Using Modified Effort" (Modified Proctor Test—ASTM-D-1557). According to these procedures, a soil sample is dried, pulverized, and separated into two fractions by a No. 4 sieve. Approximately 6 pounds of the finer fraction is moistened with a quantity of water and mixed to uniformity.

The soil is then placed in a steel cylinder in lifts and compacted with a weight which is allowed to fall a measured distance a specified number of times. When the cylinder is filled, the wet unit weight is determined, and a sample of the soil removed for moisture determination.

When testing field soils against laboratory data, it is necessary to attempt to maintain the moisture content in the field near optimum. If the soil is wetter or dryer, the compactive effort necessary to reach 95 or 100% of maximum is greatly increased. Wet soil should be dried, dry soil should be wetted.

The unit weight of soil in the field may be checked by several procedures. A classical method involves excavating a small hole within a compacted lift, measuring its volume, determining the wet density of the soil removed, and then determining the moisture content. Comparison of the dry density with the laboratory determined value results in the "percent compaction."

A relatively new technique used to determine soil density is a meter which employs nuclear technology. In this equipment, a source of radiation is inserted into a small hole to a measured depth (approximately 6 inches) into the compacted layer. The transmission of certain radiation back to a monitor through the soil is proportionate to its density. In the same instrument, a fast neutron source emits neutrons and measures the scatter of neutrons that is proportionate to the moisture content.

4.6   FROST HEAVE

When water freezes, it expands 9% in volume. When partially saturated sand or gravel are frozen, the expansion of the water is absorbed into the air-filled pore spaces and little disruption of the soil structure occurs. However, when an essentially saturated fine grained soil (silt or clay) is frozen, ice lenses are formed roughly parallel the freezing surface. Thickness of the individual lenses may increase to a thickness of several centimeters. The overall impact is that the surface elevation of the soil may raise significantly, including any structure or equipment which is resting on the surface. Other fixtures which are founded below the "frost line" may not be affected. Connections of piping, concrete, and similar structures which extend between frost-raised structures and nonmoving structures may rupture or separate.

During the freezing process the ice crystals force the soil grains apart and cause an increase in the void ratio. When thaw occurs, the soil structure is "open" and filled with water. The resulting soil is often in the plastic or liquid range and considered "mud," which severely limits field work.

4.7   STRUCTURAL FAILURE OF SOILS

Consideration of how soils respond to changes applied to them during investigations and remediation activities is the main soil structural concern to the environmental specialist. When a soil is unable to support the applied load, it can be considered a failure. Structural failure can occur as a result of overloading a foundation, which causes undue settlement, collapse of the sides of an excavation, or slope failure on the sides of a hill, dike, or similar feature. This section presents a discussion of some of the characteristics of soil failure, with emphasis on those types of failure which occur in a short-term period.

4.7.1   Slope Stability

Probably the most instances of soil structural failure encountered by environmental specialists are those of slope failure. These collapses usually result from construction of soil slopes which are too steep.

Trench excavations are often made at remediation sites for purposes of investigation, drainage, utility line placement, and slurry wall construction. The depth to which these excavations can be made without side wall failure is dependent upon the type of soil, cohesion, angle of internal friction, and water content.

Excavations into cohesionless soils (dry sand or gravel) will not support vertical slopes for any period of time (see Figure 4.10). Temporarily (very short-term) the side walls may appear to be stable, but that stability is the result of "artificial cohesion," which is the result of adhesion effect of moisture between the soil grains. An analogy to this situation is the ability of wet beach sand to be used to construct sand castles. As soon as the sand dries, the structure collapses. The stable slope of a sand trench wall is almost always less than the angle Φ for that soil (see Table 4.2).

Excavations into cohesive clay soils can theoretically proceed to a calculated depth while sustaining vertical walls. Figure 4.10b describes an excavation into clay soil. Cohesion between the soil particles provides a tension force which supports the walls until the weight of the wedge of adjacent soil exceeds the strength of the cohesion. At that depth, collapse is likely to occur at any time, without warning. The weight of the soil wedge also includes the weight of any overlying fixtures, such as trucks or construction materials setting near the trench.

The inherent instability of trench excavations is a major cause of construction accidents. Even the most carefully conducted laboratory testing cannot mirror differences of the physical structure of the site soil. No soil is perfectly homogeneous or isotropic; therefore, failures of trench walls should be anticipated and prevented when possible. With the safety of field crews in mind, the

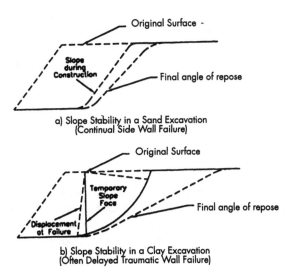

a) Slope Stability in a Sand Excavation
(Continual Side Wall Failure)

b) Slope Stability in a Clay Excavation
(Often Delayed Traumatic Wall Failure)

Figure 4.10   Slope failures during excavation.

Table 4.3   Maximum Safe Side Slopes in Excavations

Soil Type	Side Slope (Vertical to Horizontal)	Side Slope (Degrees from Horizontal)
A	¾:1	53°
B	1:1	45°
C	1½:1	34°

Occupational Safety and Health Administration (OSHA) has prepared guidelines for side slopes on excavations.

For excavations less than 20 feet deep, side sloping is a standard method of preventing cave-ins. Sloping means that the sides of an excavation are laid back to maximum safe angle from which they are not likely to collapse. Side slopes are calculated as the angle of the slope measured vertical to horizontal, i.e., a 1:2 slope indicates that the horizontal length of the slope is twice that of the vertical.

When an excavation is made, a "competent person" should evaluate the soil to determine the proper side slope. Some general soil classification guidelines may be used to estimate slope stability.

Type A Soil: Cohesive soils which are very stable, have an unconfined compressive strength of greater than 1.5 tons/ft^2, are not fissured, previously disturbed, and do not contain dipping beds. Examples include: clay, silty clay, clay loam.

Type B Soil: Cohesive soil which has an unconfined compressive strength of 0.5 to 1.5 tons/ft^2.

Type C Soil: Low strength cohesive (unconfined compressive strength less than 0.5 tons/ft^2) or granular soils such as sand or silty sand.

Reasonably safe side slopes for each of these soil types is presented in Table 4.3. The slopes presented in this table are only estimates; if there is any question concerning the safety of a slope, a professional soils engineer should be consulted.

EXAMPLE 4.1

Problem: An undisturbed sample of saturated clay weighs 1528 grams in its natural state and 1055 grams after drying. Determine the natural water content. If the unit weight (density) of the solid constituents is 2.70 gm/cm^3, what is the void ratio, porosity, and natural weight (kg/m^3 and pounds/ft^3)?

Solution:

(a) Moisture Content

$$\begin{aligned} \text{Wet Weight} &= 1528 \text{ g} \\ \text{Dry Weight} &= 1055 \text{ g} \\ \text{Water Content} &= 473 \text{ g} \end{aligned}$$

$$\% \text{ moisture} = \frac{\text{Weight of Water}}{\text{Dry Weight of Soil}}$$

$$\text{Moisture} = \frac{473 \text{ g}}{1055 \text{ gm}} = .448 = 44.8\%$$

(b) Void Ratio (e)

Since the soil is saturated, all the voids are water filled. Therefore, the volume of voids is equal to the volumetric water content.

$$e = \frac{V_v}{V_{\text{Total}} - V_v}$$

Volume of Voids (V_v) = 473 gm water = 473 cm³

$$\text{Volume of Solids} = \frac{1055 \text{ gm}}{2.70 \text{ gm/cm}^3} = 3.91 \text{ cm}^3$$

Total volume = 473 cm³ + 391 cm³ = 864 cm³

$$e = \frac{473 \text{ cm}^3}{864 \text{ cm}^3 - 473 \text{ cm}^3} = 1.21$$

(c) Porosity (n)

$$n = \frac{V_v}{V_{\text{Total}}} = \frac{473 \text{ cm}^3}{864 \text{ cm}^3} = .547$$

(d) Natural Unit Weight (Metric)

$$1 \text{ M}^3 = 1 \times 10^6 \text{ cm}^3 \quad 1 \text{ kilogram} = 1000 \text{ gm}$$

$$\text{Volume} = \frac{864 \text{ cm}^3}{1 \times 10^6 \text{ cm}^3 / \text{M}^3} = 8.64^{-4} \text{M}^3$$

Total Weight of Sample = 1528 gm = 1.528 kg

$$\frac{1.528 \text{ kg}}{8.64 \times 10^{-4} \text{M}^3} = 1.769 \text{ kg/M}^3$$

(e) Natural Unit Weight (American)

$$\text{Weight of Water} = \frac{473 \text{ gm}}{454 \text{ gm/lb}} = 1.04 \text{ lb}$$

$$\text{Volume of Water} = \frac{1.04 \text{ lb}}{62.4 \text{ lb} / \text{ft}^3 = 0.0167 \text{ ft}^3}$$

$$\text{Weight of Dry Soil} = \frac{1055 \text{ gm}}{454 \text{ gm/lb}} = 2.32 \text{ lb}$$

$$\text{Volume of Dry Soil} = \frac{2.32 \text{ lb}}{2.7 \times 62.4 \text{ lb/ft}^3} = .0138 \text{ ft}^3$$

Total Volume of Soil = $.0167 \text{ ft}^3 + .0138 \text{ ft}^3 = 0.0305 \text{ ft}^3$

$$\text{Sample Weight} = \frac{1528 \text{ gm}}{454 \text{ gm} / \text{lb}} = 3.37 \text{ lb}$$

$$\text{Unit Weight} = \frac{3.37 \text{ lb}}{0.0305 \text{ ft}^3} = 110.5 \text{ lb} / \text{ft}^3$$

EXAMPLE 4.2

Problem: A natural soil has a weight of 129.3 g and a volume of 56.5 cm³. After drying, the sample weighed 121.4 g. If the unit weight (density) of the solid particles is 2.7 g/cm³, calculate: (a) water content, (b) void ratio, and (c) percent saturation.

(a) Moisture Content

$$
\begin{aligned}
\text{Wet Weight} &= 129.3 \text{ g} \\
\text{Dry Weight} &= 121.4 \text{ g} \\
\text{Water Content} &= 7.9 \text{ g}
\end{aligned}
$$

$$
\text{Water Content} = \frac{7.9 \text{g}}{121.4 \text{g}} = .065 = 6.5\%
$$

(b) Void Ratio (e)

$$
e = \frac{V_{voids}}{V_{total} - V_{voids}}
$$

$$
\text{Volume of Solids} = \frac{121.4 \text{g}}{2.7 \text{g/cm}^3} = 44.96 \text{cm}^3
$$

$$
\text{Void Volume} = 56.50 \text{cm}^3 - 44.96 \text{cm}^3 = 11.54 \text{cm}^3
$$

$$
e = \frac{11.53 \text{cm}^3}{56.50 \text{cm}^3 - 11.54 \text{cm}^3} = 0.26
$$

(c) Percent Saturation (S)

$$
\text{Saturation} = \frac{\text{Volume of Water}}{\text{Volume of Voids}}
$$

$$
S = \frac{7.9 \text{cm}^3}{11.54 \text{cm}^3} = 68\%
$$

## CHAPTER 5

Soil Physics

5.1   INTRODUCTION

Soil physics is the study of the physical processes which are active in the wide range of materials defined as soil. These processes are the factors which make soils "work" in the sense that soils sustain plant life, support buildings, store and conduct water, act as waste disposal systems, and are an inexpensive construction material for roads and dams. Indeed, soils are very important to the existence of life on earth as we know it.

Other chapters in this text focus on the specific aspects related to the chemical, biological, and bulk structural properties of a soil mass. This chapter focuses on the physical interactions of soil grains, water, organic matter, and soil gases, and presents a discussion of how they work together in a complex environment. No soil is a homogeneous, isotropic body which is infinite in its extent. The only universal property of soil is that all soils tend toward "dynamic equilibrium"; that is, all of the factors acting on a particular soil, in an established environment, at a specified time, are working from some state of imbalance to achieve a balance.

This chapter is organized to present the physical factors which are essential to the understanding of soil functions. The first section discusses water in its basic forms. Succeeding sections present soil particle organization (structure), and fluid (water and air) storage in and transportation through the soil. The final section discusses briefly the regimen of heat flow and soil temperature.

5.2   WATER

One of the most ubiquitous materials on the earth, water is the only one which can exist simultaneously as a solid, liquid, and gas. The properties which allow water to exist in all these states also are responsible for its intricate interaction with soil materials. Compared to other common fluids, it has high melting and boiling points, heat of fusion, heat of vaporization, specific heat, dielectric constant, viscosity, and surface tension.

The specific heat of liquid water is 1 cal/degree per gram @ 15°C. In comparison, ice has a value of 0.5, iron, 0.106; mercury, 0.033; air, 0.17; and dry soil, 0.2. Water's heat of vaporization is 540 calories per gram, while that of methanol is 263; ethanol, 204; acetone, 125; benzene, 94; and chloroform, 59. All of these organic liquids have molecular weights which are greater than water. The high heats of vaporization in a light molecule suggest the presence of strong bonding forces.

In its liquid state, a water molecule has a diameter of 3 angstroms (3×10^{-8} cm). One cubic centimeter of water contains 3×10^{22} molecules. Three isotopes of both hydrogen and oxygen are found in nature; however, for introductory purposes, we may assume that hydrogen has an atomic weight of 1 and oxygen 16. The center of an oxygen atom has eight protons and eight neutrons, and is surrounded by eight electrons. Six of the electrons are in the outer orbits, with a valence of -2. Hydrogen has only one proton in its nucleus and one electron.

The water molecule has a structure of H-O-H; however, the structure is not in a straight line. Figure 5.1 demonstrates that the two hydrogen atoms are at an angle of approximately 104° from each other. This arrangement of atoms results in a slightly polar charge distribution. The "dipole" distribution of charge is similar to that of a magnet. One end exhibits a slight positive polarity, while the other is negative. The overall molecule has no net charge, but the distribution of

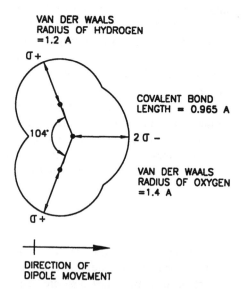

Figure 5.1 Model of a water molecule. The curved lines represent borders at which van der Waals attractions are counterbalanced by repulsive forces.

charges causes local poles on the molecule and allows water to be an efficient solvent, assists in its attraction to other molecules, and in its hydration of other ions and colloids.

The primary bond between the oxygen and hydrogen is relatively strong. Pure water has only a slight tendency to dissociate into ions. The dissolution constant for water is 1×10^{-14}; this means that in each liter of water, 1×10^{-7} moles of H^+ and 1×10^{-7} of OH^- are present. This relationship may be expressed as:

$$[OH^-] = [H^+] = 1 \times 10^{-7} \text{ moles per liter}$$

The concentration of each variety of ion (OH or H) multiplied by the concentration of the other always equals 1×10^{-14}. If the OH^- is present at 1×10^{-10} moles per liter, then the concentration of H^+ will be 1×10^{-4}.

A more convenient form of this relationship is that of pH, which is expressed as:

$$pH = -\log_{10} [H^+]$$

Values of pH less than 7 (H^+ ions dominate) are considered acidic; greater than 7, basic. It must be stressed that differences between one pH value and the next are not linear, but a function of 10. At pH 5, H^+ ions are present at ten times the number at pH 6.

5.2.1   Hydrogen Bonding

While the bonds between H and O within the same molecule are very strong, there is also a tendency of hydrogen to be attracted to neighboring molecules. This tendency is called hydrogen bonding. While this secondary link is not as strong as the primary, it exerts a significant influence on the properties of water. Water may be considered a polymer, in that each molecule of water has a significant attraction to all of its neighbors (Figure 5.2). This bonding reaches its maximum in ice. In liquid water, hydrogen bonding is responsible for surface tension, viscosity, and increased solvency.

As liquid water cools, the random vibration energy of the molecules decreases and they come closer together. As the distance becomes smaller, the density increases until it reaches a maximum of 1 gram per cubic centimeter at 4°C (39°F). Below this temperature, the forces of hydrogen bonding increase sufficiently to start the process of crystal formation. As the molecules organize, the 105^0 angle of the H-O-H structure is stressed into a hexagonal form, with a void in the middle (Figure 5.3). The expansion resulting from crystal formation (with a void in the distorted structure) causes a decrease in bulk density to 0.9 g/cm³ (thus ice floats). The force exerted by this expansion has been measured at values ranging to 30,000 psi, sufficient to fracture the strongest rocks.

The environmental significance of frozen water floating can be appreciated by observation of ponds, lakes, and rivers when ice forms on the top. If ice were

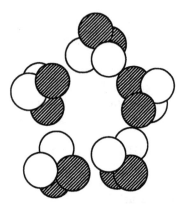

Figure 5.2 Hydrogen bonding: the formation of liquid water through hydrogen bonding of randomly oriented water molecules.

heavier than water, as most solids are in relation to their liquids, aquatic life would be forced upward to its wintry demise.

During the freezing process, the orientation of molecules into crystal form releases energy at the rate of 80 calories per gram. When cooled to 0°C, no further drop in temperature will occur until all of the water has formed into ice. After it is frozen, the temperature will continue to lower as additional heat is removed. When ice is heated, it absorbs the same quantity of heat that it lost upon freezing. (Consider ice cubes in a glass of beverage.)

At the other end of the spectrum, the vapor state, hydrogen bonding resists the escape of water molecules into the vapor phase. Breaking of hydrogen bonding requires a net heat infusion of 540 calories/gram at 100°C (at a pressure of 1 atmosphere). At lower temperatures, the heat required to release water from the liquid state is greater; evaporation at 25°C requires 580 calories/gram. The quantity of heat required to cause vaporization at less than atmospheric pressure is lower because the reduced number of air molecules in the rarified atmosphere offers escaping molecules less resistance. The reverse is true at pressures greater

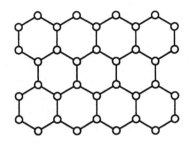

Figure 5.3 The crystalline structure of ice.

than atmospheric. When water is condensed from vapor to liquid, the same quantity of heat is released.

When water sublimes; that is, passes directly from vapor to solid (or the reverse), the "latent" heat involved is the sum of that required (released) for melting and vaporization. (That is a minimum of $80 + 540 = 620$ cal/g.)

5.2.2   Solvent Properties

Water has often been termed the "universal solvent" because it is capable of dissolving at least a small concentration of almost all chemicals. This unique capability is based on several molecular characteristics, including the dipolar nature of the molecule, dissolution of water into O and OH ions, and the attraction of hydrogen and oxygen molecules to other molecules.

The dipolar and ionic nature of water is responsible for the availability of + and − charges. Many ionic substances (such as salts) are very soluble in water because they are capable of dissociating into separate charges and "reacting" with the water. A simple example of this reaction is that of common sodium chloride:

$$Na^+Cl^- + H^+OH^- = NaOH + HCl$$

The reactions described above are possible because water has a high "dielectric constant," which means that the water molecules tend to surround the other polar molecules and prevent them from uniting with their original partners (Figure 5.4).

A second type of reaction is hydrolysis, which is the result of a metal reacting with water:

$$Na^o + H^+OH^- = Na^+OH^- + H^o$$

The metal (Na) becomes a hydroxide and the hydrogen escapes as a gas. The resulting solution becomes alkaline (basic) because of the increase in the OH^- ion concentration.

Water also is capable of dissolving non-ionic (uncharged) chemicals because of the attraction of O and H atoms in the water molecule to neighboring molecules. Several examples of this solution formation are described in Figure 5.5. Most non-ionic molecules are organic (carbon and hydrogen compounds), and all have some degree of solubility. This is extremely important in environmental studies and remediation design.

When water contains solutes, the dissolved substances interact with the water molecules to depress the freezing point and to increase the boiling temperature. Solutes, whether ionic or nonpolar, are dispersed between water molecules to prevent unrestricted association with other water molecules. Table 5.1 presents the effects of dissolved sodium chloride on the temperatures of freezing and boiling of water.

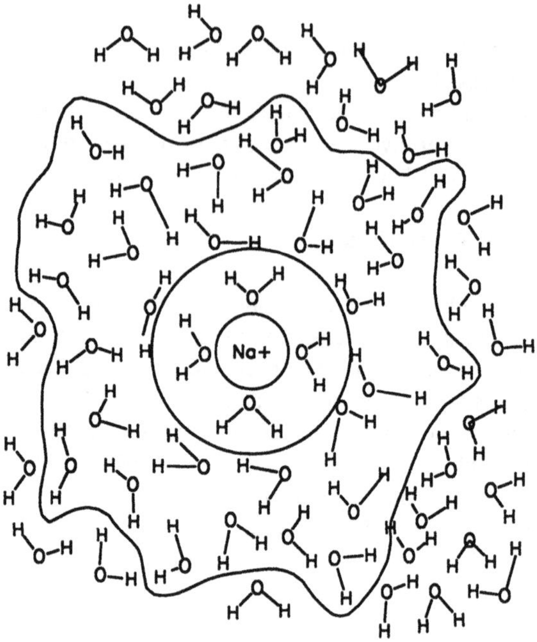

Figure 5.4   A model of the hydration "atmosphere" of sodium ion; an inner shell of more or less rigidly structured water surrounded by a cluster of looser but still structure-enhanced water, the whole floating in a sea of "free" water. (Based on Hillel, D., *Introduction to Soil Physics*, Academic Press, San Diego, CA, 1980.)

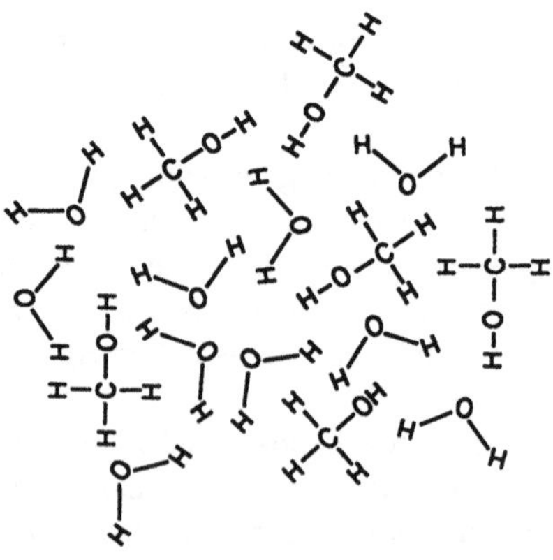

Figure 5.5   Non-ionic solutions: the tendency of attraction between the O and H, and the repulsion between like atoms sustain the solution.

Table 5.1   Effect of Sodium Chloride on Water Freezing

% NaCl	Freeze Point Depression Depression °C
1	0.6
5	3.0
10	6.6
15	10.9

5.2.3   Physical Properties

Density and Viscosity

The physical properties (principally for soil physics: density, viscosity, and surface tension) of water are also a function of its chemical structure.

Density is a measure of the mass per unit volume. The magnitude of density is dependent on the number of water molecules which occupy the space of a unit volume. As the internal energy (measured by temperature) increases or decreases, the molecules vibrate more or less frequently and strongly, and thus change the distance between themselves; this expands or diminishes the volume occupied by a given number of molecules.

As discussed previously, liquid water reaches its maximum density at 4°C, and its minimum at 100°C. Table 5.2 lists the density and viscosity of water at several temperatures. For most common purposes related to soils work, the density may be considered to be a unit weight (1 gram/cubic centimeter or 62.4 pounds per cubic foot).

Hydrogen bonding within the water mass causes the water to resist deformation (or shearing) by a property referred to as viscosity. Viscosity is the capacity of a fluid to convert kinetic energy (energy of motion) into heat energy. This property is the result of cohesion between fluid particles and also to

Table 5.2   Physical Properties of Water

Temperature °C	Density gm/mL	Absolute Viscosity	
		Centipoise	lb-sec/ft^2
0	0.9999	1.792	0.374×10^{-4}
4	1.0000	1.567	0.327×10^{-4}
10	0.9998	1.308	0.272×10^{-4}
20	0.9982	1.005	0.209×10^{-4}
30	0.9957	0.8807	0.183×10^{-4}
50	0.9881	0.5494	0.114×10^{-4}
70	0.9778	0.4061	0.084×10^{-4}
100	0.9584	0.2838	0.059×10^{-4}

interchange of molecules between layers of different viscosities. Fluids of high viscosity flow slowly, while low viscosity fluids flow freely.

Mathematically, the relationship between shear stress and viscosity may be expressed as:

$$\tau = \mu \frac{\Delta V}{\Delta Y}$$

Where:
  τ = force necessary to cause displacement
  V = velocity of displacement
  ΔY = thickness of fluid
  ΔV = difference in velocity
  μ = viscosity

which is an expression of the proportionality between viscous shear resistance and the rate of change of velocity in the direction perpendicular to the shear stress. Figure 5.6 is a graphical presentation of this relationship. An analogy to this expression may be considered a deck of playing cards pushed sideways, parallel to the face of the cards.

The factor of proportionality is called "absolute viscosity," and is usually measured in centipoise (which is 1/100 poise). The formal definition of a poise is "one gram per centimeter-second." As a convenience, the viscosity of water is equal to one centipoise at 20.2°C. Table 5.2 presents the viscosity of water at different temperatures.

Application of viscosity calculations in environmental soils work can be appreciated by the following example. As viscosity is a measure of the resistance to flow, the rate of water infiltration into the surface of the ground from irrigation or rainfall reflects its viscosity value. At 4°C, the viscosity of water is 1.57 centipoise; at 30°C, the value is 0.88 centipoise, approximately half. Therefore, during a hot summer day, the infiltration rate would be almost twice that of a cool winter day.

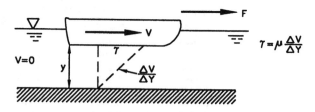

Figure 5.6   The interpretation of Newton's law of viscosity.

Surface Tension and Capillarity

It is a common observation that water rises in clays and fine silts (and, to some extent, in other soils). This rise is termed capillarity, and its cause is complex. Two of the primary factors of capillary rise are surface tension and adhesion. Surface tension is a property of the water alone; capillarity is a property of the interaction of the water with a solid.

Hydrogen bonding between water molecules is active in three dimensions. Within a mass of water, each molecule is attracted to all others which surround it. However, at the surface of a water body, the water is bounded above by the air. At this interface, the upper layer of molecules is not in contact with liquid water on one side, and the opportunity for bonding is restricted. In order to satisfy its attractions within this restricted opportunity, the outer layer molecules are forced to bond in essentially two dimensions, which increases the bond-strength along the surface, creating a layer of increased molecular attraction. Expressed in other terms, surface tension is the added cohesion of water molecules on the surface. While this layer of outer molecules is only on the magnitude of 1×10^{-6} mm thick, it has a profound influence on the behavior of water in a porous medium. These relationships are presented on Figure 5.7. Table 5.3 is a listing of values of surface tension of pure water at common temperatures.

The ultimate expression of surface tension is the raindrop. In a regime of no effective gravity and where friction is not a factor, raindrops are spherical. This form is the result of water assuming the shape which has the least surface area per unit volume. Should the volume be decreased, the sphere becomes smaller and the surface tension increases as the degree of curvature becomes greater.

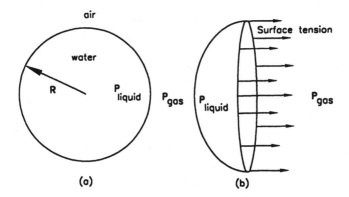

Figure 5.7   (a) A spherical droplet of water in air. (b) A force balance on the left hemisphere of the droplet.

Table 5.3   Surface Tension of Water

Temperature	Surface Tension	
°C	g/cm	lb/ft
0	0.0756	0.00518
10	0.0742	0.00508
20	0.0727	0.00497
30	0.0711	0.00486
40	0.0695	0.00475

Small amounts of electrolytes (salts or ionic materials) dissolved in water increase the surface tension. In groundwater, dissolved minerals act as electrolytes which increase the tension values. Conversely, small quantities of organic substances such as soap, alcohol, and acids decrease surface tension. The tension-suppression effects of these organic compounds "makes water wetter" and makes it possible to stretch a water film while blowing bubbles.

The second important factor leading to capillarity is that of adhesion of water molecules to solid mineral surfaces. Mineral compounds all contain elements which have some attraction to either the hydrogen or the oxygen atoms of the water molecule. The most common of these compounds are the many varieties of silicate minerals. The basic silicate structure is that of the silica tetrahedra or octahedron of silicon dioxide (SiO_2). The oxygen in the silica molecule provides an opportunity for hydrogen bonding. An example of this bonding can be observed when raindrops adhere to a glass window pane. Only when the drops become large (heavy) enough, does gravity overcome the adhesion and the drops slide to the bottom.

Solids which have a positive attraction to water are said to be hydrophilic and surfaces which repel water are hydrophobic. When drops of water are placed on surfaces, the interaction can be observed at the contact between the drop and the surface. A stationary water drop on a flat surface tries to hold its spherical shape; however, the forces of gravity and hydrogen bonding (or lack thereof) attempt to spread the water over the surface. Surfaces which attract water force the water into a shape which has a low contact angle with the surface; a high contact angle occurs when the surface repels water. Figure 5.8 demonstrates the relationship between attracting and repelling surfaces.

A paraffin-water contact angle is commonly 107°, which indicates that paraffin is a good water repellant. Quartz and many other minerals have contact angles of less than 90°, which indicates that they "wet" well with water. Mineral soil particles have typical contact angles which are significantly less than 90°, and the adhesive force is so great that removal of all of the water can be accomplished only by evaporation (not by any flow of water).

Capillary rise in small tubes (or soil pores) is the result of the combined action of surface tension and adhesion. Figure 5.9 depicts the conditions which result in the water rise. By adhesion, water attempts to cover all of the tube wall;

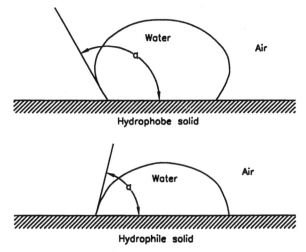

Figure 5.8 Hydrophile solid.

however, surface tension in the molecular layer in contact with the solid surface
cannot exceed the weight and internal cohesion of the water. The water molecules in the surface layer are joined to interior molecules by cohesive forces. As
the adhesive force between the water and the tube wall draws the water upward,
the surface tension drags the surface film upward, raising the column of water
against the force of gravity. The limit of this upward movement is reached when
the supporting capacity of the surface film is reached. Because the water in the

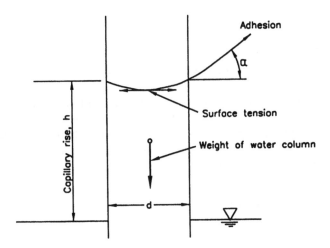

Figure 5.9 Causes of capillary action.

Table 5.4   Capillary Rise in Various Soils

Soil Type	Average Grain Size in Millimeters	Capillary Rise in Meters
Sand	2–0.5	0.03–0.1
	0.5–0.2	0.1–0.3
	0.2–0.1	0.3–1.0
Silt	0.1–.05	1.0–3.0
	0.05–0.02	
Silt-Clay	0.02–0.006	3–10
	0.006–0.002	10–30
Clay	<.002	30–300

After Testa, S. M., *Geological Aspects of Hazardous Waste Management*, Lewis Publishers, Boca Raton, FL, 1993.

column is under tension, the water pressure in the capillary tube is below atmospheric pressure (negative pressure).

The major contributing factor to the height of capillary rise is the value of surface tension; as stated previously, the smaller the size of the water drop, the greater the surface tension. The same analogy represents capillary rise. Capillary rise is inversely proportional to the diameter of the tube (or pore size of soil). Therefore, finer soils have higher capillary fringes. Table 5.4 presents the height to which water will rise by capillary forces in several soil types. The idealized equation for calculation of capillary rise is presented below:

$$h = \frac{\pi \sigma \cos \alpha}{d \gamma}$$

Where:
  d = diameter of the tube
  γ = unit weight of water
  σ = surface tension
  α = contact angle between water and the tube.

The above equation is also applicable to the capillary rise of other liquids. Where nonaqueous fluids are present in subsurface soils at contaminated sites, several cases of capillary fringes containing petroleum liquids have been documented. Typically, the heights of these fringes are less than that to which water would rise because of the weaker surface tension.

5.3   SOIL "SOLID PHASE"

The solid particles of soil consist of granules or grains of different sizes and shapes which are composed of a wide variety of minerals and organic matter.

The size of an individual grain can range from too small to see with anything except an electron microscope, to those which can be easily seen by the naked eye. Very few soils are composed of grains of all the same size, and most are heterogeneous in their vertical and horizontal extents. The shapes of individual grains are the result of the complex processes of chemical activity, mineralogical structure, climatic environment, and mechanical abrasion.

The natural occurrence of soils almost always exhibits a vertical "profile" which reflects the soil formation. Profiles range widely from highly layered features of transported (by water or wind) soils to those which formed in place as a result of weathering from underlying bedrock. The one common feature is that the upper portions tend to be more highly weathered and adapted to the climate and vegetation present.

Figure 5.10 presents a classic profile of a soil formed in place. This hypothetical profile is not a particular soil, as the diversity of soils is so great that none is identical to another. However, the form indicates the differences of appearance and structure.

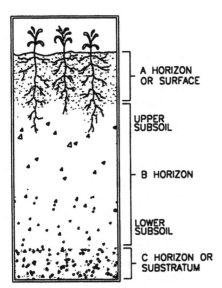

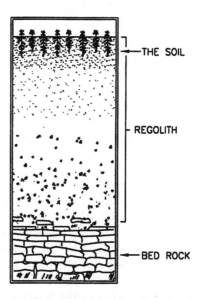

Figure 5.10   (a) A representative soil profile. The A horizon is relatively high in organic matter and becomes the furrow-slice when the soil is plowed. The B horizon, while markedly weathered, usually contains but little organic matter. At variable depths it merges gradually into the C horizon.

(b) Diagram showing the relative positions of the regolith, its soil, and the underlying rock. Sometimes the regolith is so thin that it has been changed entirely to soil, which in such a case rests directly on bedrock.

At the top of the profile, the "O" horizon is the accumulation of partially de-composed plant residues. The presence and/or thickness of this layer is deter-mined by land use, climate, and vegetation. The actual top of the soil profile is the "A" horizon, which is the zone of major biological activity, and strongest leaching by percolating water. This zone tends to be the darkest color because of the increased organic content. Below the "A" is the "B" horizon, which ac-cumulates the minerals (often clay and carbonates) leached from above; this ac-cumulation results from a change in acidity and oxygenation with depth. Un-derlying is the C horizon, which is the partially weathered parent material or bedrock. Unconsolidated material in the C horizon often grades into solid bedrock.

In humid climates, the A horizon tends to become acid due to the concen-tration of organic anions. A common feature in arid climates is the development of a "pan" (caliche) in the B horizon. Pans develop as limited rainfall leaches minerals from the upper layers and transports them into the B horizon. Howev-er, since the quantity of water is limited, the water is consumed by wetting of the soils, leaving none to transport the minerals to a greater depth. At the depth where the water stops migrating, a layer of minerals (usually carbonates) is deposited.

5.3.1 Soil Structure

In the soil classification discussion of Chapter 2, most of the emphasis was based on the categorization of discrete samples of soil for physical properties. This type of microscale analysis is suitable for individual masses of soil which are being evaluated for specific characteristics. To be meaningful, the soil sam-ple used for classification must be homogeneous and representative. Environ-mental studies and remediation projects usually are concerned with relatively large masses of soil, often including multiple horizons, and different geological strata.

In a given natural environment, soils are not often homogeneous; most soils have a macrostructure which affects the chemical properties. Macrostructure in soil often includes aggregated grains which are a composite of several (or many) individual grains. These "peds" (aggregate of grains) may be weakly cemented by chemical bonding or moisture cohesion, or organic-matter binding. During formation, some soils retain partial reflection of a former geological structure, or individual grains may be oriented by depositional processes. Roots, worm holes, cracks, bedding plains, or thin layers of nontypical minerals may also ap-pear within an otherwise uniform soil.

The composite effect of macrostructure is that bulk soils often exhibit dif-ferent characteristics than particle-size distribution or unit weight would indi-cate. Soils containing significant clay content often have greater permeability than expected. This phenomenon may be the result of the clay having been "floc-culated" (precipitated) onto the surface of other mineral grains, which tends to create aggregated grains, with open space between the particles for flow of water. A second example is the case where the clay soils were deposited in thin layers with distinct bedding planes; water flow across these bedding planes can be much greater than a simple lab test would indicate. Microstructure accounts for the majority of flow through clay or clay-dominated materials.

At the other end of the aggregated grain spectrum are the soils which have less permeability than expected. Some soils have been exposed to weathering conditions which have caused precipitation of minerals into pore spaces and on the surface of individual grains. Other porous soils have had an influx of very fine organic matter which was flushed in by rainwater moving through cracks, root holes, or connected large pore spaces and blocking them.

Careful field observation and testing are critical to the understanding of every specific site location. Undisturbed samples provide the best representation for observation and testing. Recommended procedures for sampling and analysis are presented in Chapter 8.

5.3.2 Soil and Water Relationships

The quantity of water contained in a soil mass and how it is contained is an important factor of the soil's response to its environment. Water in soil is responsible for plant growth, transport of nutrients, distribution of pollutants, and recharge of drinking water aquifers. Soil water is also responsible for physical and chemical properties such as plasticity, shear strength, cohesion, shrinking/swelling, and soil gas exchange.

Several factors are important to evaluating water in soil; among these factors are: soil wetness (moisture content), free energy per unit mass (potential), and pressure (matric potential). Each of these is discussed in the following sections.

Soil Wetness

The terms "soil wetness" and "soil moisture" are often used synonymously to describe the quantity of water found in soil. Either term may be used, as long as the user defines the application. Water in soil may be attached to the surface of soil grains or contained within mineral structures. The method used to determine the water content must be suitable for the information sought. Most often, water content is determined on a weight to weight (water to soil) basis; however, the same data may be applied in reporting procedures as volume to volume.

Oven Dry Method

The procedure used more than any other is oven drying. This test method uses the weight to weight method by drying a measured quantity of soil at 105° C for 24 hours. Moisture content then is calculated as:

$$W = \frac{\text{dry weight of soil}}{\text{weight of water}}$$

where W = water content (expressed as %).

While this method is a standard procedure and is universally recognized, three cautions should be observed:

(a) Some clay grains may not be totally dry after 24 hours because of strong adhesion of water to the mineral grains.
(b) Chemically bound water loosely held *within* clay crystal lattice structures may be removed by heating.
(c) Some organic matter (carbon or hydrocarbon) may be oxidized, resulting in lost weight.

Use of the oven drying procedure, while effective, is a destructive procedure; i.e., it requires a new sample for each test. If the purpose of the testing is to determine how the moisture content changes with time or pressure or any other parameter, a new sample must be recovered for testing. Collection of the new sample may require disruption of the soil volume being tested to render it nonrepresentative of the soil mass. Two alternative, nondestructive test procedures are discussed below.

Electrical Conductivity

Electrical conductivity of soil is directly affected by the soil's water content, texture, and proportion of soluble salts. When the texture and soluble salt content are constant, then the moisture content can be determined by changes in electrical conductivity. This procedure is especially effective for monitoring changes in moisture content during infiltration or evapotranspiration studies, or when determining irrigation cycles.

Instruments used to measure electrical resistance are often porous blocks with imbedded dual conductor wires (Figure 5.11). These blocks are buried in the soil

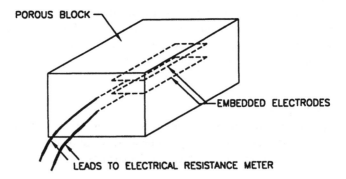

Figure 5.11   An electrical resistance block. The embedded electrodes may be plates, screens, or wires in a parallel or in a concentric arrangement.

and connected to an ohm meter. When properly calibrated at moisture equilibrium, they provide reasonably accurate data on moisture content. Blocks may be manufactured of fiberglass, plaster of paris, or porous ceramic material.

Neutron Scattering

Recent developments of nuclear technology (within the past 25 years) have produced a procedure for measuring soil moisture content by neutron scattering. This procedure is rapid, nondestructive, requires low intensity of labor, and is repeatable. Basically, a probe containing a source of fast neutrons and a detector are lowered into a prepared access hole; gauge measurements are then recorded. Ratios of neutron counts indicate the moisture content. The probe can be reinserted at leisure, and additional measurements made.

The neutron source is a mixture of a radioactive emitter of alpha particles (helium nuclei) and beryllium. Many commercial units use a two to five millicurie pelletized mixture of radium and beryllium, which emits about 1600 neutrons per second per milligram (or millicurie) of radium. Emitted source energies vary from 1 to 15 MeV (million electron volts) and an average speed of 1600 KM per second (thus "fast neutrons").

As the fast neutrons travel into the soil, they collide with various atomic nuclei, creating a spherical zone of influence around the probe. With each collision and resulting scatter the neutrons lose some of their energy. With lower energy their speed decreases until it approaches a value which is characteristic for particles at ambient soil temperatures (approximately 2.7 kilometers per second at an energy of 0.03 eV). The neutrons then become "slow neutrons." The adjacent sensor is calibrated to measure the intensity of returned slow neutrons.

The nuclei most commonly encountered in soil which have the closest mass to neutrons are hydrogen, which are then the most effective neutron moderators. If the soil contains sufficient hydrogen, the fast neutrons collide near the source and the rate of return of slow neutrons is greatest. The ratio of the source emissions to return rate can be calibrated to indicate soil moisture content.

Certain elements (boron, cadmium, and chlorine) tend to absorb slow neutrons. Tests in soils containing large concentrations of these elements may yield erroneous results. Fortunately, most soils are good candidates for this technology.

5.3.3   Energy States of Water in Soil

Movement of water in the subsurface is caused by differences in energy, always from areas of higher to lower energy levels, in an effort to reach a state of static equilibrium. Both kinetic and potential energy are operative; however, the rate of groundwater flow is so low that the kinetic energy is negligible.

Potential energy is a measure of the amount of work a body can perform by virtue of the energy stored in it relative to its surrounding environment. Understanding of the relative potential energy at each specific point can assist in evaluation of the forces acting in all directions, and thus determine if flow will occur.

The baseline for comparison of energy levels is a reservoir of free-phase water at a specified temperature, elevation, and atmospheric pressure.

Saturated soil below the water table is considered to be in a state of positive potential energy because the force of gravity is dominant. Given the opportunity, water below the water table will flow by gravity. Above the water table, in unsaturated soil, water molecules are confronted by the attractive forces of the surface of each soil grain and surface tension at the contact between the water surface and the air above. The tendency for movement in the unsaturated zone is upward until the combined upward forces are balanced by gravity. Water held at equilibrium above the water table is considered in a state of negative potential (or soil suction). The potential which results from soil grain interaction with water is often called "matric" potential.

Soil Moisture Characteristic

In a given soil profile, an energy equilibrium (albeit temporary) develops between the soil and water. Below the water table, essentially all of the pores are water-filled and the potential energy increases with depth. Above the water table, capillary forces sustain an almost saturated condition up to the height at which the capillary lifting force equals the downward force of gravity. Above the capillary fringe, some water is retained surrounding each soil grain and between adjacent grains. The funicular zone contains sufficient water to spread across the necks of some of the pores, while the pendular zone has only sufficient water to coat individual soil grains.

Figure 5.12 is a graphical presentation of a typical soil profile at an equilibrium of water holding capacity. The horizontal axis expresses the degree of saturation, while the vertical axis indicates the relative potential energy of each zone.

Potential energy of soil may be expressed in a variety of terms. When expressed as energy per mass, the units are ergs per gram or joules per kilogram, in dimensions of L^2T^{-2}. When expressed as energy per unit volume the units are dynes per square centimeter, Newtons per square meter, bars, or atmospheres in dimensions of $ML^{-1}T^{-2}$. Potential energy in soil may also be expressed in terms of energy per unit weight or hydraulic pressure. When expressed as a column of water, one atmosphere (1 bar) has an equivalent water hydraulic head of 1033 cm (or 760 mm of mercury). In this text, the standard units are those of hydraulic pressure expressed as soil suction (inverse or negative pressure).

Relative holding capacities of moisture may be related to easily observable phenomena. At low soil suctions (0-1 bar) the water held at equilibrium is under the effect of capillary forces. At higher suctions (lower pressures), the remaining water is held by adhesion and specific surface attraction of the soil grains. At a suction of approximately 15 bars, plant roots are no longer able to extract water from the soil. Ultimate "drying" of soil can only be accomplished by use of high energy processes (such as addition of heat) to liberate the remaining water.

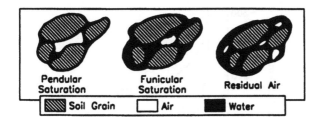

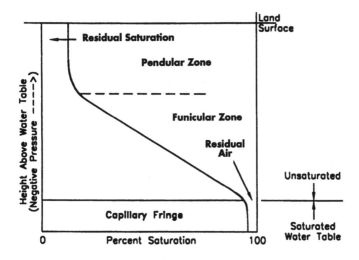

Figure 5.12   Soil zones defined by water saturation. (After Abdul, 1988.)

Hysteresis

The matric potential (suction) of a soil at a particular moisture content is not a uniform parameter. The potential of a soil which is in the process of drying is greater than the potential of a dry soil which is being wetted. Figure 5.13 shows a typical relationship between moisture content and moisture potential (expressed as a negative value). The difference between the wetting and drying paths of the curves is known as hysteresis. Causes of hysteresis are:

(a) Pore spaces of a soil tend to be nonuniform in size and shape. Migration of water into or from one pore to another; the radius of the meniscus changes, creating the "ink bottle effect" shown on Figure 5.14.
(b) The contact angle of a meniscus advancing over dry solid pore walls is greater than that of a receding liquid edge (Figure 5.15).
(c) Entrapped air (itself a fluid) must be displaced from pore spaces before water (as the second fluid) can enter them.
(d) The pore volume and shape of the matrix may be altered by shrinking and swelling of clay minerals with water retention.

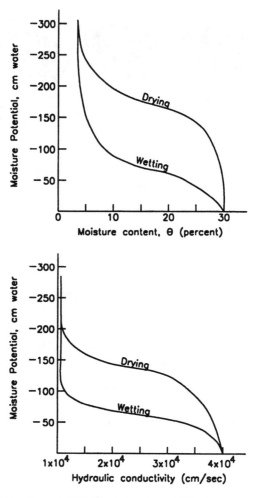

Figure 5.13 Graphs showing similarities between moisture content vs. moisture potential and hydraulic conductivity vs. moisture potential. (After Fetter, 1980.)

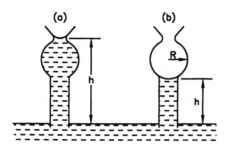

Figure 5.14 "Ink bottle" effect determines equilibrium height of water in a variable-width pore: (a) in capillary drainage (desorption) and (b) in capillary rise (sorption).

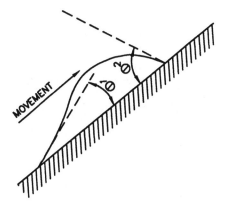

Figure 5.15   Liquid-solid interface contact showing advancing and retreating contact angle.

The net effect of hysteresis for soils having the same moisture content is that soils which were previously dry and are currently being wetted to that moisture have a lower soil suction. Alternatively, as that moisture content of a previously wet soil is being reduced, the capillary binding of the remaining water becomes progressively stronger. This increase in soil suction is caused by the remaining pore spaces becoming progressively smaller with smaller radii of curvature at the pore necks. Also, the matric attraction to the pore walls is greater because the water molecules surrounding the grains are closer to the mineral surface. Van der Waals forces increase dramatically near the mineral wall surface.

Measurements of soil moisture potential can be made by a tensiometer, as shown in Figure 5.16. A basic tensiometer consists of a porous cup (usually constructed of ceramic or Teflon) which is buried in the unsaturated zone and connected to the surface by a tube. Pore spaces in the cup are sufficiently small to restrict water flow under normal atmospheric pressure. Fine-grained silica flour is placed around the cup to assure direct connection with surrounding soil grains. The annular space around the surface tubing is sealed with bentonite slurry or similar sealant. At the surface, a vacuum gauge (or manometer) is attached to measure difference of pressure between the inside and outside environment. When the tensiometer is filled with water, soil suction draws water from the porous cup into the soil until the vacuum in the tensiometer equals the suction of the soil. Gauge readings at the surface then indicate the negative potential of the soil.

When left in the ground for extended periods of time, tensiometers indicate changes in the matric suction of the soil water. As recharge or drainage occurs, the tensiometer responds. Because the physical contact between the cup and the surrounding soil is seldom perfect, there is a time lag between the changes in soil moisture and the gauge response. Tensiometers are limited to a maximum

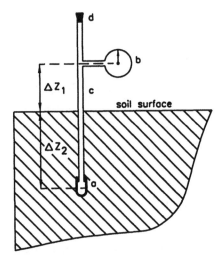

Figure 5.16   Tensiometer with vacuum gauge. (a) porous cup; (b) vacuum gauge; (c) connecting tube; (d) removable rubber stopper for deaeration.

potential measurement of 1 atmosphere suction. While the 0.8 to 1.0 bar operating range may seem to appear restrictive, it encompasses the greatest quantity of available water. Free-phase water cannot exist for any extended time at suction (negative or inverse pressures) of less than 0.7 bar (above this suction, at lower total pressure, it evaporates).

Tensiometers are useful to measure water seepage through unsaturated soil, to determine when irrigation is necessary, and to assist with observations of moisture-soil relationships, especially near the capillary fringe.

Advances in electronic science have resulted in the development of thermocouple psychrometers to measure soil moisture potential by electrical procedures. Once installed, these accurate and durable instruments allow reliable monitoring with a minimum of field effort.

Water Flow in Saturated Soils

Fluid flow through narrow tubes (or pore spaces) is affected by the fluid's physical properties of viscosity and density in direct proportion to the pressure applied. Viscosity can be conceptualized as resistance to shear. Figure 5.17 illustrates the resistance which occurs between two closely spaced plates which have a liquid between them. Some liquid adheres to each plate and as the top plate moves, the velocity distribution between the two plates is linear. The fluid velocity is proportional to the distance from the lower plate. If a constant velocity difference is to be maintained between the two plates, a steady horizontal force must be applied to the top plate, as shown on Figure 5.6.

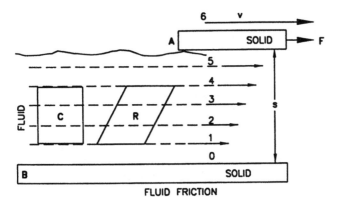

Figure 5.17   Fluid friction.

A similar condition exists when a liquid is passed through a narrow passageway or pore space. As the liquid adheres to the tube walls, a velocity gradient is developed, as shown on Figure 5.18.

The velocity gradient distribution is applicable to laminar flow conditions where the fluid molecules flow parallel to the conduit surface. Under conditions of greater velocity, the kinetic flow energy becomes sufficiently great to partially overpower the attraction of the water molecules to the conduit wall. At these higher velocities, eddies develop along the walls, resulting in turbulent flow conditions. The end result of turbulent flow is that the flow velocity is no longer proportional to the pressure drop. Fortunately, the driving force causing flow of water and other liquids through soil is almost always sufficiently small (under natural conditions) to assure laminar flow.

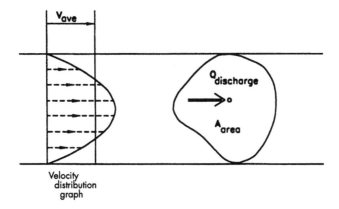

Figure 5.18   The average velocity concept.

Darcy's Law

In a saturated soil mass where the connected pore spaces can be compared to a very large number of parallel tubes, the overall flow rate through the soil is proportional to:

  (a) the open area of the cross-section, and
  (b) the pressure head applied and the length of the flow path.

These assumptions are valid only on the "macroscale" of a mass of soil which is sufficiently large that the effect of a single pore space is statistically small compared to the sum effects of all the pore spaces. Flow through individual pore spaces will not be uniform due to the presence of necks between individual grains and void spaces.

The rate of water flow through a cross-sectional area of a soil mass is equal to the equation:

$$Q = KIA$$

where:
  Q = Quantity of water discharged from a cross-section in a specified time,
  K = Hydraulic Conductivity (permeability of the matrix related to water),
  I = Ratio of drop in hydraulic head to the length of flow channel and,
  A = Area of the cross-section of soil being considered.

The term hydraulic conductivity (K) is preferred over the term permeability when describing water flow, as it is based on the physical properties specific to water. Figure 5.19 presents examples of Darcy's law applied to water flow. This basic equation is applicable when the flux (flow rate) is uniform over time, and the pressure drop over the length of the flow system is linear. If either of these conditions are violated, then a separate calculation must be made for each segment where those valid conditions exist.

The unit discharge (Q) is a flux density (or simply, flux) or volume of water which passes through a cross-sectional plane (perpendicular to the flow path) in a given time. Often Q is expressed in $L^3L^{-2}T^{-1}$ or LT^{-1}, such as cm/sec. For convenience, this statement is reduced to cm/sec. The total flux (volume of flow) must pass through only the connected pore spaces which comprise only a small percentage of the cross-sectional area. Thus, the linear flow rate (true forward velocity) of the water through the pore spaces must be greatly increased to meet the discharge flux. The following example illustrates the calculation of actual forward velocity.

Site Conditions:
  Hydraulic Conductivity (K) = 57 ft^3/ft^2 per day (or ft/day)
  Hydraulic Gradient   (I) = 0005 (slope of the water table)
  Effective Porosity   (η) = 0.12

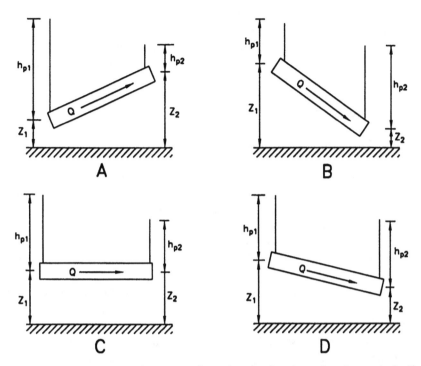

Figure 5.19 Apparatus to demonstrate how changing the slope of a pipe packed with sand will change the components of elevation, Z, and pressure, h_{p1} heads. The direction of flow, Q, is indicated by the arrow.

$$V = \frac{KI}{\eta}$$

$$V = \frac{(57)\,(.0005)}{.12} = 0.24 \ \text{ft / day}$$

Intrinsic Permeability

The ability of a fluid (any fluid, whether liquid or gas) to flow through a porous medium is dependent upon the physical properties (viscosity and density) of the fluid, on the acceleration of gravity, and on the properties of the porous media. Intrinsic permeability (k) is the basic statement of permeability of porous media, independent of any other physical parameter. Intrinsic permeability is related to all of the common factors as:

$$K = k\mu / \gamma g$$

where:
- K = conductivity of a specific fluid,
- k = intrinsic permeability (or specific conductivity) as cm^2 which depends only on the sizes, shapes and other geometric properties of the pores,
- μ = viscosity in poise units (dyne sec/cm^2),
- γ = fluid density (gm/cm^3), and
- g = acceleration of gravity.

Flow Through Unsaturated Soils

Water migrating through the soil profile from the surface must almost always pass through an unsaturated zone. The exception to this is the condition where saturated soils extend to the surface, such as under a pond, swamp, or river. Because of the number of variables involved, unsaturated (vadose) flow through porous media is more complex (and less easily quantified or predicted) than saturated flow. Recent advances in analytical procedures have enabled a clearer understanding of the process.

Under saturated conditions, water flow is the result of the interaction of the driving force, the geometric properties of the pores, and the length of the flow path. The same general rules apply to unsaturated soil, except the driving force (head) is positive (downward) but subject to (and distorted by) areas of sub-atmospheric pressure. Capillarity tends to draw water into areas where molecular coatings are thinner and menisci have smaller radii. This "suction" draws water along hydration films around soil grains until the drawing forces and gravity are in equilibrium. Continued downward flow cannot occur until the balance of forces is completed and then exceeded in favor of the advancing front of water. Figure 5.20 describes conditions of unsaturated water flow.

The greatest moving force occurs along a wetting front where water invades a previously dry soil. Soil suction along a wetting front can be many times that of the local gravitational head. (The gravitational head over horizontal distances of a few centimeters is usually close to zero.) Strong forces are required to cause water entry into areas where water saturation is low.

Saturated soil allows flow from one filled pore to another. When all pores are filled, hydraulic conductivity is at its maximum. Unsaturated soils have residual air-filled pore spaces which result in a correspondingly lower hydraulic conductivity. During and immediately following a rainstorm, the upper part of the soil profile is (at least temporarily) mostly saturated. As this saturated section of soil drains, the first pores to release water downward are the largest, as flow requires filled pores with a high ratio of volume to confining surfaces (to minimize resistance to flow). Any remaining air-filled pores are bypassed, and tortuosity is increased. After free drainage, coarse soils (sand and gravel) often only retain significant water at points where soil grains contact each other. Change from saturated to unsaturated conditions results in a significant reduction of hydraulic conductivity.

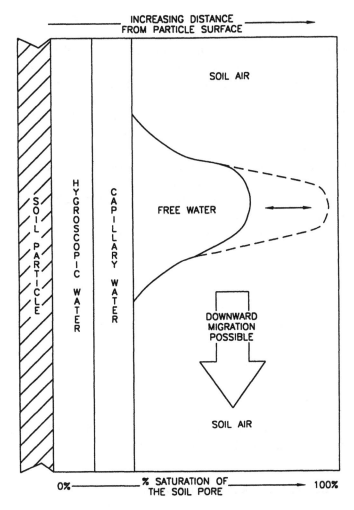

Figure 5.20   The classes of soil water. (After Dragun, J., The Soil Chemistry of Hazardous
Materials, Hazardous Materials Control Research Institute, Silver Spring, MD,
1988.)

When saturated, soils with relatively large pore spaces (i.e., sand and gravel) have hydraulic conductivity greater than clay (although clay typically has
much more pore space than gravel). The opposite is often true in unsaturated
conditions. Large pore spaces empty easily and become nonconducting by the
soil suction processes. In soil with small pore spaces (as clay and silt) many
pores retain and conduct water, even at higher suctions; thus, the hydraulic conductivity does not decrease as quickly. This phenomena becomes important in
soil profiles composed dominantly of fine-grained materials, but including layers of coarser soil. Unsaturated flow may continue downward (as an advancing
or wetting front) by suction (downward capillarity) until it encounters the coarse

soil layer. At that point, the flow ceases until sufficient gravity head is developed to force the water into the larger pore spaces.

During unsaturated flow, suction varies in relation to the soil wetness and hydraulic conductivity. When the flux is steady, the product of the flow gradient (driving head) and conductivity is constant. If the conductivity decreases, the gradient increases proportionately. When the head gradient is not linear across a section of soil, it is not possible to divide the flux by the overall gradient to determine the conductivity. The conductivity must now be considered over each individual unit.

Hysteresis, discussed previously, causes the hydraulic conductivity of a soil, at a given moisture content, to be different for wetting and drying conditions. This effect is shown graphically in Figure 5.13, which describes the relationship between hysteresis and hydraulic conductivity.

Darcy's law is applicable to unsaturated flow, when provision is made for soil suction:

$$Q = - KSIA$$

where:

-K = unsaturated hydraulic conductivity
S = soil suction
I = hydraulic gradient
A = area of flow
Q = flux

Note: It is necessary to add a - sign to differentiate downward flow from upward flow; as a convention, positive flow is downward.

5.34   Soil Atmosphere and Aeration

Unsaturated soil horizons have pore spaces which are filled or partly filled with some variety of soil atmosphere. Composition of this atmosphere is an important factor in environmental studies. Where oxygen is present in significant proportions, aerobic reactions are common, aerobic soil microorganisms are active, and roots of higher plants are able to respire. If some form of oxygen replenishment is not active or if other gases or vapors have displaced the oxygen-containing atmosphere, anaerobic conditions prevail.

Volume fractions (solid, liquid, and gas) within soils are always changing. Water and air constantly compete for the same space. Air-holding capacity varies with the soil texture in proportion to the porosity. Coarse-grained unsaturated soils (sand and gravel) commonly have porosities in the range of 25-30%. Strongly aggregated silt and clay soil, with microaggregates (>5mm in size) generally have considerable pore volume, sometimes as high as 60% or more.

Composition of Soil Atmosphere

In a well-aerated soil near the surface, the gas composition is near that of the atmosphere (nitrogen 78%, oxygen 18%, carbon dioxide 0.3%). In poorly aerated soil, root growth, microbial action, and other chemical reactions can increase the CO_2 content from the atmospheric concentration from 0.3% to as much as 3 or 4%, causing anaerobic conditions. Soil air tends to be close to 100% humidity, as water is usually present. Other soil gases commonly include nitrogen (and its anionic forms), methane (anaerobic conditions), and other products resulting from decomposition of organic matter. Where volatile organic compounds have contaminated the soil, concentrations of these products vary in proportion based on diffusion and dispersion of gaseous substances.

Convective Air Exchange

Exchange of soil air can occur by either convection or diffusion. In convection, the movement is driven by gas pressure differences. Barometric pressure changes in the atmosphere, temperature gradients, and wind gusts over the surface result in some air exchange. Other, more efficient methods are the rising and falling of the water table (to "pump" the air) and the forced ventilation of the soil as a part of remedial activities.

Theoretically, the air permeability of dry soil is approximately 50 times greater than the water permeability through saturated soil. This comparison is based on the viscosity ratio of water to air. The same general equations may be used to predict the rate of air or water movement though soil as long as appropriate measurement units are used.

Forced ventilation as a form of remedial action at chemically or biologically contaminated sites is a developing technology. The principle behind "soil venting or air sparging" relies upon the action of the increased oxygen content of the fresh air introduced into the soil. If the contaminant is a volatile compound, it will diffuse into the passing air in an attempt to reach concentration equilibrium. Compounds which are readily oxidized will react with the oxygen in the air, and aerobic microbial activity will be enhanced. Further discussion of soil venting is presented in succeeding chapters.

Diffusion Air Exchange

Diffusion of air components in soil is a result of molecular migration from areas of higher concentration to areas of lower concentration. Transfer of oxygen and carbon dioxide by diffusion can occur in either the gaseous phase or the liquid phase.

Diffusion through gas-filled pores allows oxygen (as individual molecules rather than as a mass of oxygen within a larger mass of air) to migrate from higher concentrations at the surface to areas of lesser concentrations at depth. As oxygen is consumed by biological or chemical reactions, it is replaced by diffusion. The higher concentrations of carbon dioxide produced by bioactivity diffuse toward the lower concentrations of the surface atmosphere. The rate of diffusion depends largely on the total volume and tortuosity of connected pores which are available for diffusion.

Diffusion can also be associated with the liquid phase. When oxygen-saturated rainwater moves through the soil, exchange occurs at the air-water interface. Oxygen is released and carbon dioxide is absorbed. It is commonly believed that water-transported oxygen is the primary source of replenishment to deeper subsurface. Water saturation of oxygen is typically 4–8 ppm, adequate to support aquatic life in surface water and also microorganisms in the subsurface.

5.4   HEAT TRANSFER THROUGH SOIL

Thermal balance in soil is almost always controlled by the temperature gradient. At depths below one or two feet, only in rare instances are chemical or biological activities sufficient to produce significant quantities of heat. Soil temperatures are determined by heat transfer within the soil, and by exchange of heat between the soil and the atmosphere.

Heat transfer can occur by the processes of conduction (from one particle to another), convection (by fluid motion) and radiation (by electromagnetic waves), with or without phase changes. Water, with its unique properties of latent condensation, evaporation and freezing/thawing, is a primary agent for heat transfer in the subsurface.

Heat transfer through soil has no simple relationship to the heat conduction properties of soil components; there are too many varieties of materials to allow averaging. The rate of heat transfer depends greatly on the manner in which the best conducting mineral particles are (1) mutually connected by touching and (2) separated by the poorly conducting water phase. At low water content, heat conductance is least, because heat transfer by air is minimal (even less than that of water), and heat is only transported at points where the mineral grains contact each other. Small increases in moisture content have little effect because water only coats the mineral grains. After the grains are fully water-coated, a large difference is found, as water is a much better conductor than air. Maximum heat transfer occurs at water saturation, even though the most efficient transfer of heat is between the solid grains in contact.

In undisturbed soil, temperatures change with variations in the surface temperature. Diurnal (day/night) and (seasonal) fluctuations result in soil temperature which varies in pattern to a sine wave which decreases with amplitude with depth (Figure 5.21). At some point, which depends on the soil type and moisture content, the soil temperature approaches the yearly average annual atmospheric temperature. This relationship is not entirely universal, as a few areas of the earth are "hot spots" which receive heat from volcanic activity (an increased

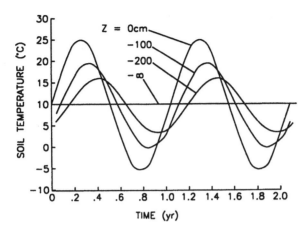

Figure 5.21 Graph of the temperature at the surface and three depths as a function of time.

geothermal gradient). These areas are relatively uncommon and are locally recognized.

During daylight hours, most incoming solar radiation is in the form of short-wave radiation, some of which is readily absorbed and some is reflected. The percentage absorbed depends on the angle of incidence, the color of the soil, vegetative cover, and the wave length of the radiation. During darkness (and densely cloud-covered days), heat is lost by longwave radiation. Net heat gain results in evaporation of water from the soil surface, transpiration of water from vegetative cover, heating of soil, and heating of the atmosphere.

When sunshine heats dry soil, as in desert areas, water evaporation consumes very little energy. Downward heat migration occurs slowly because of the low transfer properties of the dry soil, and the temperature of the surface increases greatly. After dark, the hot soil loses heat by longwave radiation to the atmosphere, causing relatively dramatic cooling. Dry soils experience very large diurnal temperature fluctuations.

Where water at the surface is plentiful, evaporation consumes considerable energy, which produces a moderating effect. Maximum evaporation is caused when wind passing over the soil produces a layer of air turbulence at the soil/air interface. A combination of freshly tilled soil, wind, and sunshine is a very efficient method of drying and cooling surface soil.

Heat loss from soil can also be caused by transpiration by plants. As solar energy is absorbed by leaves, the process of photosynthesis produces plant tissue from carbon dioxide, water, and other nutrients brought to the leaves from the root system. Water can enter the plant from anywhere in the root system and be transported to the leaves. The driving force of this water movement is the tremendous potential difference between the soil and the atmosphere via the plant tissue. This potential is the result of both osmosis and capillarity of the plant structure. Water exiting the leaves removes excess oxygen and heat, by evaporation. The net result is that the leaves are cooled to prevent overheating, and the soil is cooled by the shade of the plant.

CHAPTER **6**

Soil Chemistry

6.1 INTRODUCTION

The soil environment is a very complex setting where the various media present are constantly attempting to reach dynamic equilibrium. The processes involved include physical, biological, and chemical changes. Chemical reactions are responsible for the formation of mineral compounds, attenuation of chemicals as they move through the subsurface, release of nutrients, and a host of other factors which influence the soil environment.

Soil chemistry is too complex to be described solely by the processes of precipitation, adsorption, or exchange processes. Other complicating factors include the time scale over which these reactions occur, size, extent, and distribution of pore spaces. Groundwater may be available (or not) in the pores to distribute soluble chemicals through dispersion or transport. Water is also responsible for dissolving natural soil components, as well as those added by man. Because many of the molecules involved in these reactions are large and complex, a dynamic equilibrium is established very slowly.

Figure 6.1 shows the cycles of mass transfer between the atmosphere, biosphere, the three soil phases, unsaturated (vadose zone), and groundwater. Soils are the source of dust, and the recipient of airborne particulate matter. Rain falling through the atmosphere delivers small, but important quantities of oxygen, nitrogen, carbon dioxide, and other dissolved gases as well as some particulate matter. When water evaporates from the soil surface, solutes are left behind. Gases which constitute a small part of the atmosphere, are adsorbed by soils and plants. Certain gases (H_2O, CO_2, N_2O, and N_2) are released from soils.

The boundary between soil solids and soil solutions is not clearly defined. The attraction between soil solid particles and soil solutions is so strong that part of the soil water (ions and organic molecules) immediately adjacent to soil particles cannot be definitely assigned to either solution or solid.

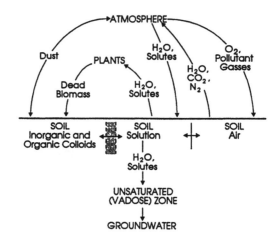

Figure 6.1   Cycle of mass transfer.

The subtle chemical interactions between soil, atmosphere, and percolating water are key components in the composition of the atmosphere and groundwater. The chemical reactions in the soil are responsible for cleansing the atmosphere and groundwater of natural or man-caused pollution.

Soil chemistry is usually divided into two divisions: reactions attributed directly to biochemical (microbial) activity, and those reactions which are abiotic (they do not depend on the presence of soil microbes). While the two branches are strongly related, the tradition of separation is continued in this text for the reader's ease. This chapter stresses the abiotic reactions, while Chapter 7 focuses on the chemical reactions attributed to microbes.

6.2   WEATHERING AND SOIL DEVELOPMENT

Rocks which were formed under the temperature and pressure conditions deep within the earth are inherently unstable when raised to or exposed at the surface. Soil which develops on these exposed rocks can be considered as a transitional phase between the minerals in the rocks and more stable chemical states. Weathering is the process of this transformation.

Weathering of igneous or metamorphic rocks alters these dense solid materials into porous unconsolidated matter whose chemical and physical composition is significantly different from the original. Weathering of sedimentary rocks is often much less dramatic, as sedimentary rocks have already been recycled (at least once) in their formational evolution.

Table 6.1 presents the composition of typical soil materials. Crystal structures, and ion valences in these materials, are stable at the conditions at which

Table 6.1   Concentration of Soil Materials

Compound	Granitic %	Basalt %	Shale %	Sandstone %	Limestone %
SiO_2	65.1	49.3	58.1	78.3	5.2
K_2O	2.4	1.2	4.3	1.4	0.04
TiO_2	0.5	2.6	0.6	0.2	0.06
Al_2O_3	15.8	14.1	15.4	4.8	0.8
Fe_2O_3	1.6	3.4	4.0	1.1	0.5
FeO	2.7	9.9	2.4	0.3	—
MgO	2.2	6.4	2.4	1.2	7.9
CaO	4.7	9.7	3.1	5.5	42.6
Na_2O	3.8	2.9	1.3	0.4	0.05
H_2O	1.1	—	5.0	1.6	0.8
P_2O_5	0.1	0.5	0.17	0.08	0.04
SO_3	—	—	0.6	0.07	0.05
CO_2	—	—	2.6	5.0	41.5
Total	100	100	100	100	100

the rock formed. When exposed to the surface, the rock becomes unstable. Freezing, thawing, wetting/drying, and exposure to chemical processes caused by dissolved atmospheric gases and biological activity break the rocks into smaller units. Most of the changes are caused by the new chemical conditions under the influence of water, oxygen, carbon dioxide, and organic compounds.

The major reaction process involved in chemical weathering results in the dissolving of mineral ions by water. In addition to the heat released by hydration, the dissolved ions enter a lower energy state while in solution. The lowest energy state would be established if each ion were to be dissolved in a solution with infinite dilution. However, under normal conditions ion dilution cannot proceed very far because soil water generally occurs as a thin film (only a few molecules thick) over the surface of solid particles. Once dissolved, an ion is free to combine with others to form a new, more stable, mineral. Later changes may involve transition to other minerals. Mineral changes are often stepwise, rather than continuous.

Some minerals, such as quartz (SiO_2), remain almost unweathered because the dissolution rate is very slow. Quartz particles tend to reduce to a limited size range (fine sand and silt) and thereafter remain stable. On the other hand, feldspars disappear from sand and silt-sized fractions rather quickly. Slow reactions only delay the time until unstable minerals will either dissolve or form new minerals. New solid minerals recrystallize into another mineral. More common, however, are reactions in which ions dissolved from one mineral are precipitated together with ions from another mineral to form new minerals. When only parts of a solute are precipitated, the process is called incongruent dissolution. Congruent dissolution is complete dissolution without reprecipitation.

Ions remaining in solution during weathering processes are rather easily leached from soil. Chemical states of ions may change during weathering, making them less soluble and thus more resistant to further weathering. An example of this process is the initial weathering of the feldspar mineral albite. This mineral is unstable at surface temperature and pressure. When the first drop of water lands on an albite crystal, the weathering is rapid:

$$NaAlSi_3O_8 + 4\ HOH + 4\ H^+ = Na^+ + Al^{3+} + 3\ Si(OH)_4$$

albite   soluble silica

The next ions to dissolve find resistance to their reaction because the first ions are present in the water droplet. The whole process is a system of equilibrium reactions that are affected by energy, solubility, and ion activity. Under conditions described above, further weathering of albite will probably result in formation of kaolinite (a clay mineral). When the water film containing any unreacted soluble ions is displaced by another drop of water, further weathering of the albite crystal continues.

$$Al^{3+} + Si(OH)_4 + 1/2\ HOH = 3\ H^+ + 1/2\ Al_2Si_2O_5(OH)_4$$

Weathering of albite illustrates several points:

1. Initial weathering releases considerable alkali and alkaline earth cations.
2. Weathering releases significant quantities of soluble silica into solution.
3. Weathering produces alkalinity in the first stage.
4. Second-stage weathering produces acidity (H^+).

Figure 6.2 presents an idealized progress of weathering in a soil profile. Basic and acidic zones migrate downward through the profile. The soil profiles shown

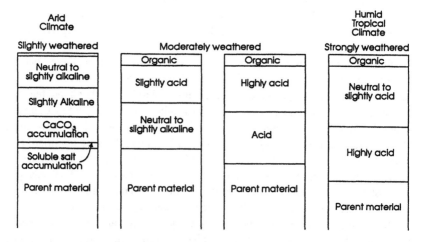

Figure 6.2   Idealized soil weathering sequences.

Table 6.2   Composition of Soils of Increasing Maturity

Compound	% Average of Igneous Rocks	% Barnes Loam (South Dakota)	% Cecil Sandy Clay Loam (N. Carolina)	% Columbia Clay (Costa Rica)
SiO_2	60	77	80	26
Al_2O_3	16	13	13	49
Fe_2O_3	7	4	5	20
TiO_2	1	0.6	1	3
MnO	0.1	0.2	0.3	0.4
CaO	5	2	0.2	0.3
MgO	4	1	<0.1	0.7
K_2O	3	2	0.6	0.1
Na_2O	4	1	0.2	0.3
P_2O_5	0.3	0.2	0.2	0.4
SO_3	0.1	0.1	-	0.3
Total	100.5	100.9	100.6	100.4

in Figure 6.2 are typical of those found in arid (slightly weathered), humid (moderately weathered), and humid (highly weathered) climates.

Secondary minerals formed by ions and molecules tend to be small and poorly crystallized. The tiny crystals of the new aluminosilicates, and Al^{3+} and Fe^{3+} hydroxy oxides have very large surface areas and many unsatisfied chemical bonds. Under this condition, they are susceptible to rapid reactions with many types of ions.

An important result of the unsatisfied bonds is that secondary minerals absorb ions to balance their charges. As most of the surface charges are negative, they tend to absorb cations.

Table 6.2 presents the composition of several soils of increasing maturity and an average composition of igneous rocks. Notice the relationship between the more and less soluble components.

The result of the previous discussion is the development of the soil profile which is encountered at each specific site. Until a site is "developed," many of the same processes continue to be active. After man's influence is exerted, the order of processes may be altered. Percolating waters may be diverted, new chemicals may be added, and biological diversity certainly will be altered. In essence, a new base for energy equilibrium will be established.

6.3   FIXATION OF ELEMENTS IN SOIL

Mobile elements (as ions) migrating with soil water are often retained by soil particles. The processes responsible for retention may be categorized into those which incorporate the element into the crystal structure of the mineral or attach it to the surface (or near surface) of soil grains. Either of these processes can be responsible for "fixation." Three general types of chemical reactions cause fixation:

1. Chemisorption: formation of a covalent bond between an adsorbed element and a mineral surface.
2. Solid state diffusion: irreversible penetration of an element into the pore spaces of a mineral's surface.
3. Precipitation: formation of an insoluble solid composed of previously dissolved material.

Fixation infers that the soil can bind and hold elements under normal environmental conditions. If the environmental conditions change, some of the element may be returned to solution.

A relatively few elements are responsible for 99% of soil and rock reactions: C, H, O, Al, Ca, Fe, K, Mg, Mn, Na, P, S, Ti, and Si. The remaining elements are very rare and are significant only on a local basis.

Elements contained in soil other than those contained in the mineral skeleton are distributed according to the following equation:

$$C_{Total} = C_{Fixed} + C_{Adsorbed} + C_{Water}$$

where:

C_{Fixed} = concentration of immobile attached elements
C_{Water} = concentration dissolved in water
$C_{Adsorbed}$= concentration adsorbed to soil surface

Note: C_{Water} and $C_{Adsorbed}$ are potentially mobile fractions

The total concentration (C_{Total}) of any specific element will not be uniform with depth in a soil profile, even under near perfect conditions. The distribution may be the result of leaching, precipitation, mobilization of trace elements due to mineral breakdown, physical breakdown of soil particles, or the mechanical flushing of clay minerals to different soil horizons. Other concentration variations in soil are caused by accumulation of relatively soluble elements on soil particle surfaces (i.e., B, Ca, and Na in arid regions), mobilization, or fixation by bioaccumulation and surface enrichment due to selective trace element uptake by plants.

Determination of potentially fixed cations cannot be evaluated or estimated based on "total" analysis for a specific ion. When laboratory analysis is made for total lead, iron, barium, or any other cation, a sample of the soil mass is dissolved and the resulting liquid is analyzed. Results of this type of analysis produce no indication of what status the ion held in the original soil.

Analysis of a clean, uncontaminated (background) soil sample for total ions will indicate the proportions of the elements present, but will not provide information on the element loading capacity of the soil.

Many varieties of specific laboratory procedures have been developed to determine the quantity of fixed, adsorbed, and dissolved ions. Most of these test procedures were developed for a specific ion. No universal test is possible because each type of soil or ion represents a unique situation.

Particular caution is recommended against attempting to use the Toxic Characteristic Leaching Procedure (TCLP) to estimate the fixed or adsorbed elements. While this procedure does evaluate the potential mobility of specific ions, the leaching fluid is acidic, which tends to dissolve some of the minerals. The TCLP is much too severe to produce useable results for fixation studies.

6.4   ION FIXATION IN CLAY AND MINERAL SOILS

Some compounds such as HCl and H_2O are miscible in all proportions and so do not result in precipitation or separation into a separate phase. Most compounds, however, have limited solubility, especially in water. When the compound has reached its solubility limit (i.e., it is saturated), precipitation occurs. A common example of precipitation is:

$$Mg^{2+} + CO_3^{2-} \rightleftarrows MgCO_3 \text{ (s)}$$

The magnesium and carbonate ions remain in solution until the solubility limit for magnesium is reached. If additional magnesium or carbonate ions are added, precipitation occurs. If the environmental setting changes, such as the system is heated, or additional water is added, the precipitated mineral has a tendency to dissolve back into solution.

If another ion is added to the solution, direct substitution may occur, especially if the valence and ion size are similar.

$$Mg^{2+} + Ni^{2+} + CO_3^{2-} \rightleftarrows (Mg,Ni)CO_3 \text{ (s)}$$

Substitution can also occur between ions of different valence and similar size, if the "coordination number" is identical. The term coordination number is the number of ions or molecules that can surround an ion in a crystal structure. The structure of the crystal lattice and other ions present have a significant effect on the coordination number. Mg^{2+} and Fe^{2+} can substitute for Al^{3+} where Al^{3+} is in sixfold coordination with oxygen. Three ions with valence of +2 exchange for two ions with a valence of +3. In a similar manner, Al^{3+} can be found in fourfold and sixfold coordination in silicate structures. It is very common to discover Al^{3+} substituting for Si^{4+} at the center of an oxygen tetrahedron. Substitution of elements is common in clay minerals and sometimes in carbonates, phosphates, sulfates, sulfides, and silicates.

6.4.1   Ion Fixation as Hydroxides (-O) and Oxyhydroxides (O-OH)

Metals in solution may combine with O and OH to form precipitates in the form of "sludges," or the combined form may be incorporated into mineral structures.

$$Al^{3+} + 6\ H_2O \rightleftarrows Al(H_2O)_6^{3+}\ \text{ or}$$

$$Fe^{3+} + 6\ H_2O \rightleftarrows Fe(H_2O)_6^{3+}$$

In combined form, each species is capable of polarizing as mononuclear, and reacting even further with water:

$$Al(H_2O)_6^{3+} \rightleftarrows [Al(OH)(H_2)_5]^{2+} + H^+$$

These mononuclear species are stable up to pH 5; above that, they form polynuclear species which precipitate as soil hydroxides, oxides, or oxyhydroxides. This same type of reaction is used in water treatment plants to encourage flocculation of suspended particles. When the hydroxides precipitate, they have very large surface areas and provide opportunity for suspended particles to attach to them and thus be removed from suspension.

Initially, these new precipitates in soils are noncrystalline (amorphous) but over extended time they may form into crystals. Also, some of the mononuclear species occasionally adsorb onto mineral surfaces as coatings or become bonding agents between grains.

6.3.2   Eh and pH Fixation

Some elements may exist at different valence states, also referred to as oxidation states. Examples of different oxidation states are $Mn^{2+,3+,4+}$ and $As^{3+,5+}$. Redox (reduction/oxidation) reactions are chemical reactions in which an element gains an electron (reduction) or loses an electron (oxidation). Electrons designated as e^- participate in chemical reactions in the same way as H^+. For example:

$$Fe^{3+} + e^- \rightleftarrows Fe^{2+}$$

Usually, the redox state is expressed as Eh, which is measured against a standard reference potential. Eh in soil is the result of a complex combination of oxidation and reduction species and reactions. Measured Eh may vary with water content, oxygen activity, and pH. As electrons (e^-) neutralize protons (H^+), many reactions are both Eh and pH dependent. Element fixation is often controlled by both Eh and pH.

The predominant species of a mineral which occurs can often be described by an Eh/pH diagram. Figure 6.3 is an Eh/pH diagram for iron. This diagram

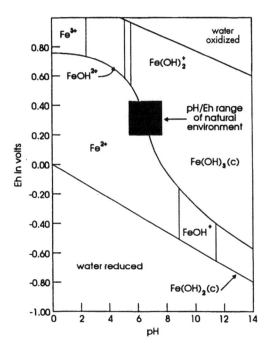

Figure 6.3   Eh-pH diagram for iron.

indicates that in poorly aerated soil (high Eh = few electrons present), and acid conditions (low pH), the independent species Fe^{3+} is the predominant form of iron. If large sources of electrons are introduced into this environment (i.e., by introduction of oxygen), the iron would be altered to Fe^{2+}. Fe^{2+} compounds tend to be much less soluble, and precipitate. This phenomena can be readily observed as red stains on plumbing fixtures where untreated high iron groundwater is used. Similar behavior is common in well-aerated soil systems which have iron oxide or hydroxide coatings on individual soil grains.

Eh/pH diagrams yield important information regarding the potential fixation of an element in soil. At different pH values, some species may form insoluble oxides, hydroxides, or oxyhydroxides. They may react readily with available sites on mineral crystal surfaces. Eh/pH diagrams enable a scientist to estimate if existing soil conditions will be conducive to fixation of added elements. The Eh or pH of a soil can be altered to promote fixation of contaminants.

The redox state of Mn, Fe, and other multistate elements often has a signif-icant influence on soil color. Bright colors (yellow, red, and reddish-brown) are common in well-oxidized soil or soil formed under oxidizing conditions. Sub-dued colors (gray or black) are indicators of reducing conditions. Soil zones which are exposed to varying conditions such as fluctuating water table may contain streaks of both gray and red-brown. These soils are referred to as "mottled" soils.

Soil colors are good indicators, but subtle differences are difficult to compare between samples. Individual investigators have varying perceptions of color. A procedure to standardize color is to use standard color chips. A Mussel color chart consists of approximately 175 color chips arranged by hue, value, and chroma. Each of these parameters is a component of color. Hue is the spectral color; red (R), yellow (Y), or red-yellow (YR). Value is the relative lightness of the color and function of the total amount of light. Value is expressed as numerical 0 (absolute blackness) to 10 (absolute white). Chroma represents the relative purity of the spectral color. Chroma increases from 0 (in neutral gray).

Written form of the Mussel color follows the order of hue, value, chroma. A typical notation is 5YR5/6.

6.4.3   Dissolution and Release

Solubility equilibrium is based on water's ability to hold ions in solution. Consider the following reaction:

$$Mg^{+2} + CO_3^{2-} \rightleftarrows MgCO_3 \text{ (s)}$$

When the concentration of Mg^{2+} is greater than the solubility of magnesium in water, precipitation occurs. The converse is also true. $MgCO_3$ (s), as well as other minerals, tend to buffer the concentration of Mg^{2+} in the soil. If environmental conditions change, the equilibrium is driven toward one side. Dissolution of minerals may result in formation of new minerals (crystalline or amorphous), or the ions may remain in solution to be carried away by the natural groundwater flow to another location. Whether new minerals form or not depends on the concentration in the soil.

The polar nature of water is responsible for the dissolution of soil minerals. Crystal faces are composed of ordered positive and negative ions. When a sufficient number of appropriately charged ions congregate around an ion of a crystal, they can overcome the forces which attach the ion to the crystal and pull that ion into solution. Ions on corners of crystals or edges of crystals are more easily removed because they have more exposed surface area.

Substances which have equal attraction to either the crystal or water (NaCl, KCl, $CaCl_2$) are easily dissolved. Substances which have greater affinity for the crystal tend to be resistant to weathering. Metal oxides and aluminosilicate minerals tend to be relatively stable (insoluble), and thus many soils are composed of these compounds.

If the available water has an Eh/pH potential which encourages dissolution, the soil minerals will be more easily weathered. Addition of acids to soils often increases dissolution. Minerals tend to dissolve if:

1. An acid is added to the soil which lowers the pH.
2. A strong base (alkaline material) is added to raise the pH.
3. Bulk quantities of an ion are added to a solution at equilibrium. If NaCl is added to a saturated solution of $MgCO_3$, the Cl will combine (complex) with Mg^{2+} to form $MgCl^+$ and $MgCl_2O$. This complex development will decrease the Mg^{2+} in solution and cause $MgCO_3$ to dissolve in an attempt to replace the Mg^{2+}.
4. Soil conditions facilitate the formation of a new mineral which is less soluble than the original mineral. For example:

$$CoCO_3 \text{ (solution)} + H_2S \rightarrow CoS \text{ (s)} + H_2O + CO_2$$

Additional Co is then dissolved to replace that removed from solution.

6.4.4   Remediation Based on Element Fixation

At locations where inorganic chemicals are spilled or leaked into soil, remediation procedures are often based on pragmatic considerations. If it is possible to control the environmental setting, fixation of the leaked chemical may be the best procedure to prevent its further migration.

An example of oxidation precipitation was employed at a site where a large quantity of K_3AsO_3 (potassium arsenate) was released into soil. The arsenic in this compound was the As^{3+} species. A solution of $Ca(OCl)_2$ was injected to convert the As^{3+} to As^{5+}.

$$2K_3AsO_3 + Ca(ClO)_2 \rightleftarrows 2K_3AsO_4 + CaCl_2$$

Then $Fe_2(SO_4)_3$ was used to transform the soluble arsonite to insoluble arsonate:

$$2K_3AsO_4 + Fe_2(SO_4)_3 \rightleftarrows 2FeAsO_4 \text{ (s)} + 3K_2SO_4$$

6.5   ION FIXATION AND MOBILITY

The process of fixation removes ions from water and immobilizes them, with a corresponding decrease in the concentration of the ion in solution. Adsorption of an ion onto the surface of a charged particle is another process which can reduce the mobility of free ions in soil.

Cation adsorption is the process by which a negatively charged surface adsorbs positive ions. Clay crystal surfaces possess a surplus of negative charges which attract cations for balance. Humus (partially decomposed organic matter) is colloidal in structure and has many unsatisfied negative charge sites. These unmatched charges are at the end of complex organic molecules which have structures similar to R-C-O-OH⁻. The end sites are capable of loosely holding cations.

When an ion in the water phase is attracted to a soil surface, it must displace another (already present) cation. For example:

$$(Soil)Mg + Cu^{2+} \rightleftarrows (Soil)Cu + Mg^{2+}$$

This reaction is known as exchange or cation exchange. The quantity of exchangeable ions in a soil is called the *ion exchange capacity*. Often it is limited to exchangeable cations, and hence is referred to as *cation exchange capacity* (CEC).

CEC is expressed as milliequivalents per 100 grams of soil. For example: a clay with a CEC of 1 meq/100 g is capable of adsorbing 1 milligram (mg) of hydrogen (or its equivalent) for every 100 grams of clay. The equivalent amount of Ca would be 20 mg because Ca has two charges and an atomic weight of 40. Ca with an atomic weight of 40 has an equivalent weight of 40/2 or 20. Therefore, 20 mg is the weight of 1 meq of Ca. Table 6.3 presents CEC values for several representative soils and clay minerals.

CEC is reversible:

$$C_s \quad \overset{K_{des}}{\underset{K_{ads}}{\rightleftharpoons}} \quad C_e$$

where:

C_s = concentration adsorbed on soil surfaces (μ/gm soil)
C_e = concentration in water (μg/mL water)
K_{des} = desorption coefficient
K_{ads} = adsorption coefficient

Table 6.3 Cation Exchange Capacity Values for Selected Soils and Clay Minerals[a]

Soils (and Minerals)	pH	CEC meq/100g
Average of agricultural soils (Netherlands)	7.0	38.3
Average of agricultural soils (California)	7.0	20.3
Mollisol (Russia)	7.0	56.1
Sodic Desert soil (California)	10.0	18.9
Clay subsoil (Alabama)	6.0	18.0
Montmorillonite (mineral)	–	92
Kaolinite (mineral)	–	3.3
Mica (mineral)	–	25
Vermiculite (mineral)	–	125

[a]Compiled from several sources.

This same equation may be expressed in a form which defines the equilibrium distribution of ions between the water phase and soil phase:

$$K_d = C_s/C_e$$

where: K_d = distribution coefficient

The adsorption capacity of a soil is slightly different for different ion species. The primary reason for differences of adsorption is that simple free ions do not exist in aqueous solutions. All dissolved ions are surrounded by water molecules. Hydrated ions have different sizes. Since electrostatic forces are involved in the accumulation on soil surfaces, the smallest ion with the greatest charge will be preferentially sorbed. For monovalent cations, the general order is:

$$Cs > Rb > K > Na > Li$$

For divalent ions:

$$Ba > Sr > Ca > Mg$$

In general, trivalent ions are adsorbed more easily than either divalent or monovalent ions.

The formation of cation complexes is another factor which has an effect on the rate and quantity of ions adsorbed by soil. Analysis of groundwater normally represents the *total* concentration of each element in solution. Many cations exist in more than one ionic form. Each form may have a different valence which results in a different mobility because of its affinity for adsorption and solubility. The process of cation combination with molecules or anions containing free electrons is coordination. The complex formed is often called a "ligand."

The stability of a ligand can be expressed as:

$$A^+ + L^- = AL \text{ or,}$$

in terms of an equilibrium reaction:

$$K = [AL]/[A^+][L^-]$$

Where each bracket represents a concentration in moles per liter. Values greater than 10^7 are considered stable.

Although a complex may have a relatively high K value, the complex may disassociate due to the presence of a cation which can form an even more stable complex. An example is Ni-DTPA (an organic complex for which $K = 10^{61}$), which is stable until subjected to the presence of Fe, which forms an even more stable Fe-DTPA ($K = {}^{29.2}$). The effects of ligands on soil surfaces in the process of cation adsorption can be classified into five separate categories.

1. When a ligand has a low affinity for both cation and the soil surface, the ligand will migrate with groundwater. Typical of these are Cl, HCO_3, SO_4, HSO_4, NO_3, HS⁻, and occasionally CO_3 and S^{2-}. Groundwater texts describe these as "conservative" ions because they tend not to be retarded and flow at approximately the same rate as groundwater.
2. A ligand may be represented by a soluble complex which is extensively adsorbed by the soil. OH_ is probably the most important ion that affects cation adsorption. In general, as the OH_ activity increases, the extent of metal ion adsorption increases.
3. Sometimes the ligand has a high affinity for the cation and forms a soluble complex, but the complex has a low attraction to soil surfaces. Complexes which are not adsorbed generally migrate with the water. These complexes may include naturally occurring organic acids, amino acids, and polyphenols with molecular weights of less than 300.
4. In a few cases, the ligand is extensively adsorbed by the soil, but the cation has a low affinity for the anion. In this case, the cation may move freely with the groundwater.
5. Some organic ligands are extensively adsorbed by the soil but also have a strong attraction for the cation. The cation in this case does not migrate.

Examples of categories 4 and 5 above are the result of relatively large organic molecule complexes with molecular weights of greater than 400. The large molecules are held to soil surfaces of clay and silt sized grains by van der Waals forces. Humus (decomposed organic matter) particles are very active in this type of fixation.

6.6   ANION ADSORPTION

Most soil particles have predominantly negative surface charges; however, some soil surfaces *also* have positive sites which attract anions. Typical anions found in soils are: phosphate, arsenate, molybdate, sulfate, borate, silicic acid, fluoride, chloride, and nitrate. Three types of soil surfaces attract anions.

1. Soils which contain a high concentration of oxides and a high pH tend to develop a negative charge on the surface due to H^+ disassociation from OH⁻ group. If the pH is low, the oxide surface develops a positive charge by attracting H^+.
2. The edges of layered silicate minerals (clay) expose positively charged groups which can attract anions.

3. Humus often has large amounts of exposed radical fragments which may be either positive or negative. Degraded organic matter has a large propensity for ion adsorption.

In most U.S. soils, anion exchange capacity is less than 5% of cation exchange capacity. Since most soils have positively charged surfaces, anions are repulsed, and sometimes migrate by dispersion at a rate faster than groundwater flow. This phenomenon is referred to as *facilitated transport*.

6.7   EQUILIBRIUM MODELS

Adsorption of ions by soil is determined largely by the equilibrium established between the ions dissolved in water and those attached to soil. This relationship can be expressed by the following equation:

$$C_t = C_s + C_e$$

where:

C_t = total amount present
C_s = concentration adsorbed on soil
C_e = concentration dissolved in water

Because these are equilibrium reactions, the quantity of cations or anions adsorbed depends on the concentration of the cation or anion in the water. When a soil is tested over a range of concentrations, a graph can be prepared to describe the relationship between that adsorbed on soil and that dissolved in water. This type of graph is an "isotherm" which may assume any of several shapes. Figure 6.4 describes the general forms of isotherms. Any chemical adsorption equilibrium relationship can be expressed by this type of graph.

6.8   ION MOBILITY IN SOIL

Mobility of elements in soil is the result of the extent of fixation (positive or negative adsorption), complex formation, and reaction kinetics (how long is required to reach equilibrium). Some general summary conclusions can be recognized:

* Cations have low mobility in silt and clay soils.
* Cations have moderate mobility in sand and loam soil.
* Anions are slowly mobile in clay and silt soils, but more mobile in other soils.

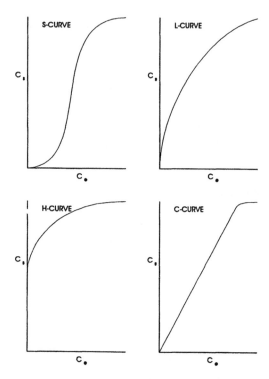

Figure 6.4 General isotherm shapes: where C_s represents the concentration of compound held on soil particles and C_c represents the concentration dissolved in the water phase.

6.9   ACIDS AND BASES IN SOILS

6.9.1   Occurrence

Ammonium ions, produced by the decomposition of plant and animal tissue by microbes, are a major natural source of H^+ ions in soils. Autotrophic bacteria oxidize ammonium ions to form H^+ and NO_{3-}. Organic acids produced by decomposition of plant residues are also important sources. In agricultural settings, acidity is increased when monocalcium phosphate hydrolyzes in water to form dicalcium phosphate and phosphoric acid:

$$Ca(H_2PO_4)_2 \overset{HOH}{\rightleftharpoons} CaHPO_4 + H_3PO_4$$

Microorganisms break down residual organic matter using reduced sulfur (S) or sulfate sulfur (SO_4^{2-}) from the soil system. Organic sulfur is reduced or incompletely oxidized into forms such as sulfides, elemental sulfur, and thiosulfates by aerobic bacteria. Oxidation of a reduced form of S releases two H^+ for each S atom oxidized:

$$H_2S + 2 O_2 \rightarrow 2 H^+ + SO_4^{2-} \text{ or,}$$

$$S + 1.5 O_2 + H_2O \rightarrow 2 H^+ + SO_4^{2-}$$

Carbon dioxide dissolves in rain water to form carbonic acid (H_2CO_3). At normal pressure and temperature, the atmosphere contains approximately 0.3% carbon dioxide, which results in rainwater with a pH of 5.6. Water in the subsurface can have a lower pH because microbial activity in soil can produce higher concentrations of CO_2 (up to 1%). The pH of 1% CO_2-water mixture is approximately 4.9.

Acid can also be produced by the weathering of pyrite in soils near mine sites and other reducing environments.

$$2 FeS_2 + 7 H_2O + 7 O_2 \rightleftharpoons 4H_2SO_4 + 2 Fe(OH)_3$$

Runoff from coal strip mine sites and spoil piles has been measured as low as pH 2. This level of pH is toxic to all forms of wildlife, and contributes to major environmental damage even when diluted.

A recent source of acid is man-made acid rain. Rain falling through air contaminated by excess quantities of NO_x and SO_4 is a major source of contamination in several areas of the world.

6.9.2   Chemical Buffering by Soil

A buffer is a solution that is resistant to change the amount of H+ present (i.e., it resists changes in pH). Soil can act as a complex buffering system as a result of the presence of organic matter, cation exchange capacity, aluminum oxides and hydroxides, aluminosilicates, and carbonate minerals.

Soil organic matter also acts as a buffer because it contains carboxyl (R-COOH) and hydroxyl (R-OH) radicals. In neutral soils these can exist in the disassociated state R-COO$^-$ and R-O$^-$. As H^+ is added, these disassociated groups preferentially absorb H^+ over other cations. The resulting R-COOH and R-OH result in the net removal of H^+ from the soil system.

Cation exchange capacity buffers the H^+ content of H_2O by adsorption to cation exchange sites on soil particle surfaces. For example, the extent of Al^{3+} hydrolysis is dependent on the Al^{3+} concentration in water. Buffering mechanisms involving metal cations in solution are affected by the extent of cation exchange sites on soil particle surfaces.

In neutral soils, solid phase aluminum occurs in the lattice structure of the interlayer regions of clay minerals. As surface concentrations of H^+ increase, the H^+ reacts with the surface.

$$(Soil)\text{-}Al\text{-}OH + H^+ \rightarrow Al^+ + HOH$$

The net result of this reaction is the creation of positively charged adsorption sites and an increase in the *anion* exchange capacity of the soil. Iron (and other metal oxides) and hydroxide minerals are believed to act as buffers in a manner similar to the aluminum minerals.

In some neutral soils, aluminosilicate minerals also act to buffer H^+. Minerals (such as allophane) and the edges of some clay minerals contain disassociated silanol groups ($\equiv SiO^-$). The H^+ reacts with the silanol, which removes it from the soil system.

Carbonate minerals act as buffers. When H^+ is added to neutral or basic soils, it reacts:

$$2\ CaCO_3 + 3\ H^+ \leftrightharpoons 2\ Ca^{2+} + HCO_{3-} + H_2CO_3$$

to release carbonate (H_2CO_3) and bicarbonate (HCO_{3-}) ions into the water phase.

In arid regions where rainfall is much less than evaporation, minerals dissolved by percolation of carbonic acid water migrate only a short distance downward. The accumulated deposits of these minerals usually occurs at shallow depths as caliche. Caliche minerals such as $NaCl$, Na_2SO_4, $CaCO_3$, and $MgCO_3$ result in *saline* soils. The pH of saline soils usually is 8.5 or higher.

Concentrations of Na salts may result in the saturation of the soils cation exchange capacity with sodium. When this occurs, a high pH may result. Hydrolysis of Na_2CO_3 produces OH^-, which increases alkalinity of the soil.

$$Na_2CO_3 + 2\ HOH \rightleftharpoons 2\ Na^+ + 2\ OH^- + H_2CO_3$$

Other metals common to evaporite salts (K, Ca, and Mg) may also act in a similar way to increase the pH.

6.9.3   Measuring Soil Acidity

Soil acidity is the most commonly measured property. Three types of acidity are represented in the soil system. The first type is solution activity, which is the amount of H^+ present in water in equilibrium with the soil. If the water is extracted, the acidity can be measured in terms of pH. General descriptions of soil acidity are shown in Table 6.4.

The pH of a soil is usually measured by mixing one part of soil with 2 parts of distilled water, and mixing. A pH probe or paper is inserted into the liquid and the results recorded.

Exchangeable acidity is the portion of total acidity that can be removed from soil cation exchange sites by equilibrium reactions with neutral, unbuffered salts (KCl, $CaCl_2$, or $NaCl$). In general, exchangeable acidity is due to Al^{3+} activity. Where organic matter is an important part of the CEC, much of the H^+ released from the organic exchange sites results from hydrolysis of aluminum.

Table 6.4 Soil Reactivity

Acid Description	pH
Extremely acid	<4.5
Very strong	4.5–5.5
Medium acid	5.6–6.0
Slightly acid	6.1–6.5
Neutral	6.6–7.3
Mildly alkaline	7.4–7.8
Medium alkaline	7.9–8.4
Strongly alkaline	8.5–9.0
Very alkaline	>9.1

The third type of soil acidity is known as "titratable" activity, which is the quantity of acidity that can be neutralized by addition of a base to the soil. Soil compounds which primarily affect titratable acidity are: (a) Al, Fe, and other metal oxides or hydroxides, (b) R-COOH and R-OH functional groups, and (c) OH^- groups attached to mineral surfaces.

6.9.4   Effects of Bulk Acids and Bases on Soil Properties

The physical effects of acids in soils varies from soil to soil, based on the quantity of acid released and the minerals present in the soil. Large volumes of low pH waste streams may alter the physical properties of soil. Addition of bulk acids may increase the quantity of free Fe, Al, and other metal cations in the soil water. H^+ is a very small ion which can easily replace Ca^{2+} and Mg^{2+}. Also, the acid may dissolve crystalline minerals. If the pH drops below 5, the soil loses its capacity to retain cations. However, if the buffering capacity is sufficient, the soil may neutralize the acid.

Addition of bulk acids may cause dissolution of clay minerals and formation of new clay minerals. In a montmorillonite clay soil, addition of bulk acids can dissolve significant amounts of SiO_2. The activity of freed SiO_2, K, and H^+ favor the formation of kaolinite.

Bulk acid additions may alter soil permeability. Acids (as HCl and H_2SO_4) favor formation of sparingly soluble soil minerals at very high concentrations. At low concentrations, mineral acids only cause minimal permeability changes.

In the presence of low concentration H_3PO_4, the buffering capacity of clay minerals tends to raise the pH to 6.8 to 8.0, which results in decreased permeability due to the blockage of pores by precipitated oxides, sulfates, and phosphate minerals.

6.9.5   Effects of Bases on Soil Properties

Large volumes of high pH material may increase the amount of cations in the water phase by dissolving more base-soluble minerals. These cations were previously fixed within the mineral structure. This effect should occur only if the soil's buffering capacity cannot compensate for additional OH⁻ and prevent the pH from rising.

Bulk addition of OH⁻ may cause alteration of the clay minerals into alternate forms. Depending on the activity of the SiO_2, K^+, Ca^{2+}, Mg^{2+}, and H^+ in the specific soil, one mineral may dissolve and another form. Permeability of most soils does not appear to be significantly affected by moderate additions of OH⁻.

6.9.6   Improvement of Alkaline or Acidic Soils

Improving the pH of soils that are more acidic or alkaline than desired is usually accomplished by neutralization. The general equation is:

$$Acid + Base \rightleftharpoons Salt + Water$$

Bases added to soil are usually considered "liming" agents because of the agricultural practice of adding lime ($CaCO_3$) or powdered limestone to soil. When $CaCO_3$ is neutralized by HCl, the reaction is:

$$CaCO_3 + 2\ HCl \rightarrow CaCl_2 + H_2O + CO_2$$

One molecule of lime can treat two molecules of acid. Addition of lime to soils is usually in granular or powdered form. Coarse particles have much less surface area and require longer time to react. Fine particles can be dispersed more widely and react quickly.

The quantity of lime material required to alter the pH of soil is a function of the acidity and the buffering capacity of the soil. Laboratory or pilot testing may be required to determine the appropriate addition rate.

Chemical compounds used to lower the pH include: elemental sulfur, sulfuric acid, aluminum sulfate, or ammonium sulfate. Aluminum sulfate is a popular horticultural agent which reacts with water and soil in two processes to lower soil pH.

$$Al_2(SO_4)_3 + 6\ H_2O \rightarrow 2\ Al(OH)_3 + 3\ H_2SO_4$$

$$6(Soil\text{-}H^+) + Al_2(SO_4)_3 \rightarrow 2(Soil\text{-}Al^+) + 3\ H_2SO_4$$

Iron sulfate acts similar to aluminum sulfate to increase acidity. Elemental sulfur is oxidized by autotrophic bacteria into acid-forming compounds.

6.10   ADSORPTION OF ORGANIC CHEMICALS ON SOILS

The fate and transport of organic chemicals is an important part of environmental chemistry. The transport of organic chemicals is dependent upon the processes of adsorption and desorption. The fate of organic chemicals in the subsurface is affected by transformation processes including: volatilization, photolysis (light degradation), hydrolysis, and biodegradation.

Organic chemicals released into soil may be retained on soil surfaces or adsorbed onto soil particles. The portion which is strongly adsorbed does not easily migrate or leach through a soil profile. Weakly adsorbed chemicals may be slowly desorbed by flowing groundwater and transported to other locations.

The extent of adsorption of an organic chemical is controlled by its concentration in the groundwater, as well as the chemical structure, and the physical and chemical makeup of the soil. Adsorption of a dissolved organic chemical by soil is an equilibrium reaction. The quantity retained by the soil is in direct proportion to that dissolved in the water.

The equilibrium coefficient may be generated by detailed laboratory testing or estimated based on the chemical's solubility in water and its octanol-water partition coefficient. Octanol is an alcohol which is almost insoluble in water. When a chemical is added to a mixture of water and octanol, the chemical will partition into each solvent at a given ratio. Estimates based on the octanol-water partition coefficient assume that water is the primary solvent.

Molecular structure determines the gross activity of an organic compound, as it is responsible for: molecular volume, water solubility, vapor pressure, and octanol-water partition coefficient. The three-dimensional structure of an organic molecule may have considerable effect on the potential for adsorption of that molecule. The larger the planer surface form, the greater the adsorption.

The size of a molecule is directly related to its probability of adsorption. The larger the molecule, the greater the tendency to be adsorbed. As the surface area of contact increases, the holding forces increase. Some very soluble large molecules are held tightly, due to their many contact points with soil particles. An example of this is soluble dye added to cotton fabric. A large portion of the soluble dye is adsorbed by the cloth fibers. Subsequent washings remove only a small portion of the color with each washing.

Recent discoveries indicate that a number of organic molecules (i.e., pesticides, polynuclear hydrocarbons, and aromatic hydrocarbons) bind easily to large dissolved organic molecules such as humic acid, fulvic acid, and organic matter. These large molecules are not mobile, as they themselves tend to be adsorbed to soil particles.

Some organic molecules prefer attachment to hydrophobic (water repelling) surfaces instead of aqueous solvents or hydrophilic (water liking) surfaces. Molecular fragments of C, H, Br, Cl, and I tend to be hydrophobic, while fragments of N, S, O, and P are primarily hydrophilic.

As a practical matter, the accumulation of an organic compound cannot be completely described by either partitioning or adsorption. Attachment of organic molecules is a complex interaction between organic chemicals, clay (or soil particle surfaces), and soil organic matter.

Some organic molecules have a built-in positive or negative charge. An example of these molecules include some metal-organic compounds. These molecules are adsorbed onto cation exchange sites.

It is important to remember that soils characteristically have more negative surfaces than positive. Some organic chemicals have different forms at varying pH. When the organic fragments have either a + or − charge, they will be attracted to the oppositely charged sites.

Hydrogen bonding occurs whenever a hydrogen atom serves as a bridge between a molecule and a possible adsorption site. The formation of a H bond as a link between a cation and a polar organic molecule is shown in this formula:

$$(Soil\text{-}cation)\text{-}OH\text{-}H\text{-}\text{-}\text{-}O=C\text{-}R$$

Another type occurs between an adsorbed organic cation and another organic molecule:

$$(Soil)\text{-}R\text{-}NH^+\text{-}\text{-}\text{-}\text{-}O=C\text{-}R$$

6.10.1 Concept of Adsorption

Adsorption of organic molecules follows the same pattern as inorganic molecules.

$$C_s \underset{K_{ads}}{\overset{K_{des}}{\rightleftharpoons}} C_e$$

where:

C_s = concentration on soil surface (µg/g)
C_e = concentration dissolved in water (µg/mL)
K_{des} = desorption
K_{ads} = adsorption

In most instances adsorption is a reversible reaction, expressed as the distribution coefficient.

$$K_d = C_s/C_e$$

Greater adsorption potential is indicated by larger K_d.

If K_d is expressed as a function of organic matter in the soil (since organic matter is often the primary adsorbent), the K_d is often expressed as K_{om} or K_{oc}.

$$K_{om} = K_d/om \text{ or } K_{oc} = K_d/oc$$

where:

K_{om} = soil adsorption coefficient based on organic matter

K_{oc} = soil adsorption coefficient based on organic carbon

om = organic matter (mg/mg soil)

oc = organic carbon content (mg/mg soil)

A commonly assumed relationship between organic matter (om) and organic content (oc) is:

$$K_{oc} = 1.724 \, K_{om}$$

In many cases where the measurement of K_{om} and/or K_{oc} is not possible, it may be necessary to estimate the value. Estimates of adsorption of non-ionic chemicals onto soil particle surfaces can be estimated reasonably well by water solubility and the water-octanol partition coefficient (K_{ow}). These estimates are useful as initial starting points. The equations listed below are approximations based on observations and several simplifying assumptions, including: the dissolved organic molecule has a molecular weight of less than 400, and the soil organic carbon is greater than 0.1%.

1. $\log K_{oc} = \log K_{ow} - 0.21$
2. $\log K_{om} = 0.52 \log K_{ow} + 0.64$
3. $\log K_{oc} = -0.55 \log S + 3.64$

(S = water solubility, mg/L)

These equations should be used with caution when the dissolved fraction of organic chemical is relatively small. They cannot be applied successfully when a free phase nonaqueous chemical is the primary solvent. Some representative K_{oc}, K_{ow}, and solubilities for common chemicals are presented on Table 6.5.

6.10.2 Predicting Organic Chemical Mobility in Soil

As dissolved organic chemicals migrate through soil, some of the chemicals are adsorbed by the soil, as discussed above. The remainder of the dissolved organic chemicals migrate with the groundwater flow. However, the concentration of dissolved chemicals decreases as the length of the flow path increases. The cause of the change in concentration is related to molecular dispersion (from areas of greater concentration to areas of less concentration, acting in three dimensions), and the tendency of the soil to retard the flow of chemical due to attraction onto soil particles (retardation).

The following equation may be used to calculate an estimate of the migration rate of a chemical in a saturated soil system.

$$V_e = V[1 + K_d(b/P_r)^{-1}]$$

Table 6.5   Water Solubility and Distribution Coefficients

Chemical	Solubility in Water (mg/L)	K_{ow} (log)	K_{oc} (log)	Potential Mobility
1,1,1-Trichloroethane	950	2.17	2.25	moderate
1,1,2,2-Tetrachloroethane	2900	2.56	1.90	high
Arochlor	0.02	6.03	4.80	immobile
Benzene	1,750	2.13	1.69	high
Aldrin	0.2	5.17	2.61	moderate
Carbon tetrachloride	785	2.64	1.36	moderate
Chlordane	1.85	2.78	4.73	immobile
DDT	0.006	3.98	5.37	immobile
Ethyl benzene	152	3.15	2.75	low
Methylene chloride	13,200	1.25	1.39	very high
Napthalene	31	3.37	3.00	low
Pentachlorophenol	14	5.01	2.95	low
Phenol	93,000	1.46	1.43	very high
Toluene	515	2.69	2.53	moderate
Vinyl chloride	1.1	0.6	ND	ND

Notes: ND = Not Determined
          Compiled from various sources

where:
  V_e = velocity of a chemical where $C/C_o = 0.50$ (concentration compared to original concentration)
  V = average linear velocity of groundwater
  K_d = distribution coefficient
  b = bulk soil density
  P_r = total soil porosity

A modification of the above equation can be used to estimate the migration of a chemical in the unsaturated zone.

$$V_e = V_{sw} (WC + bK_d)^{-1}$$

where:
  V_{sw} = amount of water percolating through the unsaturated zone (inches per year)
  WC = volumetric water content

Some general equations have been developed to provide an estimate of the convective time (t_c) which is required to move a mass of contaminant through a distance (d).

$$t_c = dbK_d + WC + P_{sa} K_h / WF$$

where:

    b   = bulk soil density
    K_d = distribution coefficient
    d   = distance
    WC = volumetric water content
    P_{sa} = volumetric air content
    K_h = Henry's law constant
    WF = water flux, quantity of water passing a line at the boundary of the system

When adsorption is relatively high ($K_d > 4 \times 10^{-3}$ m³/kg), the WC P_{sa} and K_h can be neglected. The equation can be simplified to:

$$t_c = dbK_d/WF$$

This equation will describe the movement of a front or a narrow pulse of a migrating chemical plume of organic chemicals.

6.11 ORGANIC CHEMICAL DIFFUSION AND VOLATILIZATION

The migration of volatile organic chemicals from the soil to the atmosphere is a significant concern to the environmental professional. The rate at which volatile chemicals migrate through subsurface soils and enter the aboveground atmosphere is controlled by equilibrium reactions:

Chemical adsorbed on soil particles ⇌ Chemical in soil air

Chemical in soil air ⇌ Chemical in soil water

Chemical in soil air ⇌ Chemical in atmosphere

In order for migration to occur, volatile chemicals from the subsurface must move toward the soil surface and then into the atmosphere. Diffusion is the mechanism which causes volatilization and migration from the soil to the atmosphere. Diffusion is the average rate of migration or velocity of a chemical moving through stagnant air. As in all migration, movement depends on a driving force (or gradient).

Chemicals diffuse from areas of high to areas of lower concentration. Greater differences of concentration result in more rapid movement. Diffusion can also be caused by pressure gradients, such as different density of gases. Lighter gases rise, while heavier ones sink. Gases can also diffuse according to temperature gradients. The primary method of diffusion in soils is the result of concentration gradient.

Molecular weight is a primary factor in diffusion. Nonreactive molecules usually diffuse (at a given temperature) at a given rate, which can be estimated by the following equation:

$$D_{a1}/D_{a2} = (MW_1/MW_2)^{1/2}$$

where:

D_{a1} = diffusion coefficient of chemical 1 in air
D_{a2} = diffusion coefficient of chemical 2 in air
MW_1 = molecular weight of chemical 1
MW_2 = molecular weight of chemical 2

Table 6.6 is a listing of the air diffusion coefficients of several common chemicals.

Vapor pressure is the force exerted by a chemical as its molecules attempt to escape into the vapor phase. When a liquid is sealed in a partially filled container, some of the liquid's molecules gather sufficient kinetic energy to escape into the gaseous phase. Evaporation continues until a state of equilibrium is established between the molecules in the liquid and the molecules in the gaseous state. At equilibrium, the gaseous molecules exert a pressure against the walls of the container known as the vapor pressure. The point at which sufficient en-

Table 6.6 Diffusion Coefficients for Selected Chemicals in Air

Chemical	DA (cm²/sec)	T (°C)
Ammonia	0.28	25
Benzene	0.088	25
Carbon dioxide	0.164	25
Chlorobenzene	0.075	30
Chlorotoluene	0.065	25
Diethylamine	0.105	25
Diphenyl	0.068	25
Ethyl acetate	0.089	30
Ethyl alcohol	0.119	25
Ethyl benzene	0.077	25
Ethylene dibromide	0.070	0
Ethyl ether	0.093	25
Hexane	0.080	21
Methyl alcohol	0.159	25
n-Octane	0.060	25
Oxygen	0.178	0
n-Pentane	0.071	21
Toluene	0.088	30
Water	0.220	0
Xylene	0.071	25

Based on Dragun, J., The Soil Chemistry of Hazardous Materials, Hazardous Materials Control Research Institute, Silver Spring, MD, 1988.

ergy has been added to a liquid to cause the chemical to enter the gaseous state is the boiling point.

Molecules which have a strong attraction between them have a relatively low vapor pressure. The opposite is true for weak intermolecular forces.

Van der Waal's forces between molecules and dipolar reactions influence a chemical's solubility in water and diffusion from water into air. Chemicals with a higher solubility are less likely to be released into the atmosphere. The equilibrium concentration developed between a chemical and water can be expressed as:

$$K_{wa} = (\mu g \text{ chemical/mL water}) \div (\mu g \text{ chemical/mL of air})$$

The potential for a chemical to volatilize from the water phase to the air phase can be estimated by Henry's law. This law states that, in a very dilute solution, the vapor pressure of a chemical should be proportional to its concentration.

$$V_p = K_h C$$

where:

V_p = vapor pressure
C = concentration of chemical in water
K_h = Henry's law constant

Henry's law constant may be expressed as:

$$H_h = (V_p)(MW)/760 \, S$$

where:

V_p = vapor pressure of chemical (mm Hg)
MW = molecular weight of chemical
S = solubility of chemical in water (mg/L or gm/M^3)
K_h = Henry's law constant

Important note: K_h are expressed in different dimensions throughout the literature. Units used are: atmos-m^3/mole, atmosphere-cm^3/gram, or dimensionless. Caution should be used to assure that the numerical value is based on the same units used to describe V_p, C, or S.

Generally, chemicals with a K_h less than 5×10^{-6} atmos-m^3/mole remain in the water phase. Chemicals with a K_h greater than 5×10^{-3} atmos-m^3/mole are more easily transferred to the air phase. K_h must be used with caution, because it is not valid for solutions with concentrations greater than about 3%, or if several chemicals are dissolved in the water.

Diffusion of volatile materials through soil occurs from gas-filled pore space to gas-filled pore space. Because the pore spaces are limited in size and have tortuous interconnections with other pores, diffusion is slower than in unconfined air. As the soil particles (and pore spaces) become smaller, the coefficient constant becomes smaller.

In most soils, the porosity decreases as the bulk density of the soil increases. Thus higher bulk density results in larger D_{sa}. Soil moisture can also affect D_{sa}. Where the pores are more filled with water, vapor diffusion becomes smaller. Soils with grains which adsorb the chemical also retard volatilization and vaporization. Clay minerals or dispersed organic matter both have very large surface areas and tend to retard the release of volatile chemicals.

Several climatic factors also affect diffusion and volatilization in soils. Chemical volatilization is affected by pressure gradients, atmospheric turbulence, and air temperature. When high air pressure forces air into the soil, the air mixes with the existing soil atmosphere. When the atmospheric pressure is less, the soil air vents to the surface, carrying some of the volatile chemicals. Atmospheric pressure changes occur relatively slowly and the pressure equilibrium between the soil and atmosphere is almost instantaneous; thus, the depth influenced on a daily basis is very small (less than one meter).

Atmospheric turbulence caused by the wind over the soil surface has little effect on pressure gradients, but increases volatile transfer from the soil to the atmosphere by diffusion. Continuous replenishment of "clean" air in contact with the surface prevents establishment of equilibrium concentrations, thus increasing the quantity of volatile compounds removed from the soil. Above a very thin layer near the surface, the atmospheric turbulence far exceeds the molecular diffusion rate and thus continues mass transfer of chemicals from the soil to the atmosphere.

Favorable atmospheric conditions (wind and sun) will favor the "wick effect" for enhancement of volatilization. As volatile chemicals are removed from the surface, soil water containing the chemicals continues to migrate by capillarity to the surface.

6.12 SOIL VAPOR MONITORING

Recent studies have demonstrated that monitoring of the concentration of volatile organic chemicals in the soil atmosphere can be a valuable tool in the identification of soil contamination. Soil vapor monitoring as a technique involves the collection of representative samples of the soil atmosphere which are analyzed for their chemical content.

Two general procedures are used to determine the concentration of volatile organic compounds in the soil atmosphere. The first involves the insertion of a hollow probe to a specified depth and the application of a vacuum to extract a volume of the purged gas. The purged gas is analyzed by a gas chromatograph or other analytical instrument.

An alternative procedure is to bury a vapor trapping device in the soil which adsorbs a representative sample of the vapors. This device is usually an iron wire coated with activated carbon. The sample trap is allowed to remain in the soil for a sufficient time to allow an equilibrium to develop between the vapors in the soil atmosphere and that adsorbed on the carbon. After retrieval, the sampler is taken to the laboratory where the iron wire is heated electrically until all of the vapors are released from the charcoal. These vapors are introduced into the analytical instrument (gas chromatograph or GC mass spectrometer). Resulting concentrations can be correlated to the actual field concentrations.

After receipt of analytical results from either of the above procedures, the most common practice is to plot the sample locations on a map, and contour the soil vapor data. The results should indicate the general distribution of volatile organic chemical contamination in the shallow soil.

Soil gas investigations are useful tools for site investigation; however, there are several serious limitations:

1. The flux of a chemical in soil air is the result of a concentration gradient which can exist from the concentration location up, down, or horizontal. Thus, a point measurement may not be representative of the actual plume.
2. The absence of a chemical in the soil air does not always indicate that the chemical is not present. Sometimes, the pores between the probe and source are plugged, often by percolating water, or a lower permeability layer may separate the probe from the source.
3. Downward moving water may have flushed the chemical toward the water table and suppressed the vapor transport.
4. Soil adsorption may be great, and thus the chemical may not be in the water and therefore the soil atmosphere concentration may be relatively low.

Soil Venting as a Remedial Action

The tendency for development of an equilibrium between volatile chemicals in soil air and soil can be an effective remediation procedure. Two general applications are used: the first involves the excavation of the soil and spreading over the land surface (or through a mechanical mixer). The exposure to large quantities of free air allows the contained chemicals to volatilize and disperse.

The second type of soil vapor venting employs the use of forced venting in the subsurface. The most common procedure is to install vent wells which are open to the vadose zone, and to create a slight vacuum on a selected pattern. Air from the atmosphere enters from surrounding areas (or other vent wells) and migrates toward the low pressure area in the vacuum well. As the air moves through the soil, the volatile chemicals attempt to establish equilibrium with the moving air, and remediation is accomplished.

6.13 ORGANIC CHEMICAL REACTIONS IN SOIL

Many organic chemicals in the soil environment are altered or degraded by the complexity of chemical reactions which occur in the subsurface. Some of these alterations are completed by abiotic (nonbiological) reactions, and others are the result of direct biological activity. The main nonbiological processes are discussed below:

1. Hydrolysis is a chemical reaction in which an organic chemical molecule reacts with a water molecule (or a hydroxide ion).

$$R\text{-}X + H_2O \rightleftharpoons R\text{-}OH + H^+ + X^-$$

$$R\text{-}X + OH^- \rightleftharpoons R\text{-}OH + X^-$$

where:
   R = the main part of an organic molecule (radical)
   X = functional group

   These reactions result in a broken carbon-leaving group bond, and formation of a carbon-oxygen bond.
   Some hydrolysis reactions in soil can be catalyzed by the presence of metallic ions or by clay minerals. This type of catalytic reaction can be exemplified by the hydrolysis of acetonitrile to acetamide in the presence of bentonite as a catalyst.

$$CH_3\text{-}CN + HOH \rightleftharpoons CH_3C(O)NH_2$$

2. General substitution and elimination reactions occur when one or two "groups" are extracted from a molecule and the remaining molecule portion restructures itself into a stable new molecule. The following example describes this process:

1,2-Dibromoethane   Vinyl Bromide + Hydrogen Bromide

3. Oxidation occurs when electrons are removed from a molecule (loss of electron = oxidation). The loss of an electron can be the result of several causes: a spontaneous breakdown of a molecule which contains weak covalent bonds, exposure to radiant energy (UV), disruption by high energy electrons or particles (alpha or beta), and electron transfer to an ion of a transition element contained in a clay mineral structure.
   Without the electron, the molecule becomes a free radical which is highly reactive and will unite with the first available species. This type of reaction often

involves several very complex steps, all of which are beyond the scope of this text.

4. Reduction occurs when an electron is gained (gain of electron = reduction) as a result of a chemical reaction. An example of reduction is the reduction of ethylene chloride to chloroethane.

$$H\text{-}CH{=}CH\text{-}Cl + 2H^+ \rightleftharpoons H\text{-}CH_2\text{-}CH_2\text{-}Cl$$

Reduction reactions of organic chemicals outside of biological activity are poorly understood, and not often studied. Research is needed to define the mechanisms of these processes in the context of the anaerobic environment common to groundwater settings.

5. Some organic chemicals react quickly or violently with water. If these chemicals are spilled in bulk quantity, energy produced by the reaction may be released more quickly than it can be absorbed by the local environment. General classes of highly water-reactive chemicals include: anhydrides, peroxy acids, some platinates, and phenyl-metal compounds. The reader is referred to published literature for additional information.

## CHAPTER 7

Microbiology

7.1 INTRODUCTION

This chapter presents an introduction to the basic principles of microbiological activity in soils. Specific attention is devoted to the interaction between the soil matrix, contaminants, and microbial processes as they participate in environmental restoration.

Microbes in soil are one of the primary factors that make soil what it is. The life processes of these small living organisms are responsible for many of the chemical reactions in the shallow subsurface.

Microbes, acting in soil, surface water, and groundwater recycle most of the chemical building blocks required for life. These blocks are composed primarily of carbon, nitrogen, sulfur, water, and oxygen. Based on fossil records, higher life forms were abundant on the earth over half a billion years ago. However, in geologic terms, the biological turnover of carbon, oxygen, and water are rapid. For example, it has been estimated that the carbon dioxide respired by plants and animals has an atmospheric residence time of 300 years. Oxygen in the atmosphere is recycled in about 2,000 years, and all the earth's water is split and reconstituted every 2 million years. These cycles are predominantly driven by microbial processes. Without microbial recycling, the raw materials that make life possible would have been exhausted long ago.

Throughout geologic history the ecosystem has maintained a state of dynamic equilibrium. For example, the preservation of organic carbon in the form of dead tissue has been generally balanced by the ability of microbes to transform the matter into reusable C, N, O, and H_2O. These elements, though common, have been dispersed rather evenly throughout the biosphere in concentrations that promote the continuance of life. Humans, on the other hand, tend to accumulate and concentrate chemical substances. For example, organics, such as petroleum products or municipal sewage waste, are accumulated through human

activity and presented to the environment in such a manner that the natural ecological balance is upset.

Microbes can only use a portion of the organics as a substrate or food source.
Ultimately, complex and potentially hazardous organic compounds can be degraded into benign inorganic substances. However, in many cases the biological
demand exerted by contamination exceeds microbial resources. Limiting factors,
such as nutrient availability, inhibit naturally occurring bioremediation, and the
cleanup process is arrested. Fortunately, through the use of engineering controls,
the effects of limiting factors can be mitigated. With proper design the microbiological treatment of organic waste can be highly efficient and relatively
inexpensive.

7.2 MICROBES

The term *microbe* is used to identify various forms of microscopic life. Microbes are members of the kingdom of Protista. Protists are usually divided into
two broad groups, Eucaryotes and Procaryotes. Procaryotes, considered to be the
more primitive microorganisms, include bacteria and blue-green algae (cyanobac-
teria). Procaryotes are characterized by the lack of a nucleus and absence of internal specialized organelles (proto-organs). Eucaryotes (including algae, fungi,
and protozoa) are usually many times larger than procaryotes and contain a defined nucleus and organelles.

Algae and blue-green algae can be thought of as microscopic plants that create energy through photosynthesis. Conversely, the protozoa are animal-like and
include many predatory varieties. Conceptually, fungi and some bacteria act as
scavengers. The other microbial classes do not fit well into any of these categories. Some share both animal- and plant-like qualities, while others function
in ways that are completely unique. For example, certain bacteria, found only in
deep ocean geothermal hot springs, derive energy from the oxidation of sulfur,
and obtain carbon from carbon dioxide.

Microbial types exist in various ecological niches. An important branch of
microbiology is microbial ecology, the study of microbial habitat. As one would
expect, photosynthetic bacteria (algae and cyanobacter) are essentially absent in
the subsurface. Protozoa and fungi are generally restricted to shallow soil environments. Bacteria, however, are ubiquitous in the soil, and have been demonstrated to exist even in deep geologic zones. Bacterial types, however, do vary
with depth. Bacteria capable of degrading organic matter are prevalent in the
shallow soil/groundwater environment. In deeper aquifers, where organic matter is limited, bacteria that utilize inorganic chemical reactions for energy and
carbon predominate. Although fungi serve an important function in the shallow
soil, most subsurface microbial studies emphasize bacteria because of this group's
overwhelming importance.

7.2.1   Bacteria

Bacterial types, or species, are classified by a variety of methods. Taxonomic differentiation can be based on gross morphology, motility, gram staining, growth temperature range, substrate utilization, electron acceptor type, carbon source, energy generation, spore formation, colony formation, by-product generation, enzyme synthesis, and/or other characteristics.

Bacteria may occur as individual cells or in colonies which vary in population from two to millions of cells. Individual bacteria may have shapes of rods (bacilli), spheres (cocci), or other shapes, such as spirals (see Figure 7.1). Single cells are microscopic, although colonies are often sufficiently large to be seen with a hand lens. Most individual bacteria cells are in the size range of 0.3 to 50 microns. (One micron = .001 millimeter.) Cocci are often about 2 microns in diameter, bacilli are typically about 10 microns long, and a few of the filamentous organisms reach a length of 500 microns. The smallest, such as mycoplasmas, reach only 0.135 microns.

Some varieties of bacteria have long whip-like appendages called flagella, which can cause movement. This motion can be near 0.5 feet per hour, which is remarkable, considering their size. Other bacteria have shorter appendages (pili) which assist with adherence to solid surfaces.

Bacteria are commonly classified according to energy generation and carbon source. Bacteria can be either phototrophic (capable of energy generation through photosynthesis) or chemotropic (producing energy through chemical reaction). Chemotropic bacteria are further divided into lithotrophic (energy generation through inorganic chemical reactions) or organotrophic (the generation of energy through organic oxidation). Heterotrophic bacteria utilize organic matter as

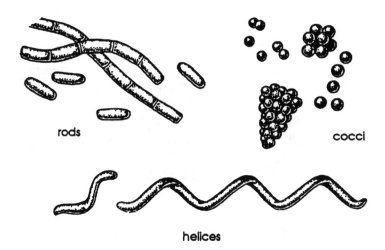

rods   cocci

helices

Figure 7.1   The various shapes the bacterial cell may have are shown in outline. These are cylindrical (rods), spherical (cocci), and helical (spirilla and vibrios).

a carbon source, whereas autotrophic bacteria obtain carbon from inorganic sources such as carbon dioxide.

The metabolic activity of bacteria is strongly influenced by their small size. The surface to volume ratio is extremely large; the inside of a cell is very accessible to chemical reactions in the surrounding medium. A parallel situation is evident in the use of finely divided catalysts in chemical reactions. The large surface area per volume encourages efficient reactions. Compared to more complex organisms, bacteria are much more efficient users of energy.

For comparison, if the surface area to volume ratio of a man is 1:20, the same ratio for a "typical" bacteria is 1:9,000,000. The importance of this comparison is that all waste products and food must pass through the bacteria's cell walls. Because they are so small, each microbe is in direct contact with its immediate environment. This intimate association with surrounding media results in a highly efficient process unit. Some bacteria even secrete exoenzymes (outside enzymes) that break down potential sources of food into soluble components which can be absorbed through the cell wall.

Compared to procaryotes, eucaryotes have a decidedly abbreviated cell morphology. Essential components include a cell wall, cytoplasmic membrane, DNA, ribosomes, and the cytoplasm. Other components are dispensable, even to cells in which they are found, and are never found in some groups of bacteria. Typical bacterial structure is shown on Figure 7.2. Some varieties are encircled by a slime layer which appears to protect the cell from attack by other microbes. The cell wall gives the cell a structure and form. The layer immediately inside the cell wall, called the cytoplasmic membrane, controls the variety of materials entering and leaving the cell. The fluid which fills the cell is cytoplasm, the

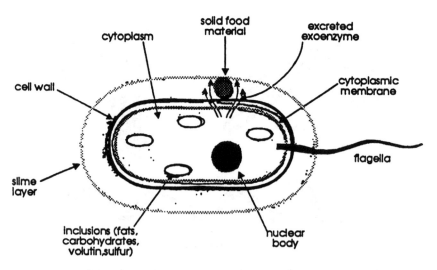

Figure 7.2 Schematic diagram of typical bacterial cell structure.

medium in which the metabolic processes are completed. A nuclear body, not a true nucleus, is the control center for metabolic processes and reproduction. In procaryotes, hereditary data for reproduction are contained in a single molecule of DNA which is located in the cytoplasm. Ribosomes are small organelles containing protein and RNA, which are the site of protein synthesis.

Many bacterial cells are able to produce "granules" during periods when abundant food or nutrients are available. These granules store essential nutrients during periods of excess. When procaryotes are exposed to rich media, they produce storage granules consisting of carbon substrate, principally polysaccharides (complex sugars).

A few species of bacteria are spore forming. Bacteria form spore bodies when environmental conditions become harsh. Spores are metabolically inactive bodies which allow survival (or resting) until conditions improve to favor normal growth. Some viable spores have been documented to have existed for several centuries.

Cell biomass measured by fresh weight is generally 75-85% water, thus the remaining 15–25% is organic or mineral matter. Of the dry weight, approximately 50% is protein, 10–15% is composed of cell wall materials, 10% is lipids, 10–20% is RNA, and 3–4% is protein. On an elemental basis, the dry weight is typically 50% carbon, 20% oxygen, 14% nitrogen, 8% hydrogen, 3% phosphorous, 1% sulfur, 1% potassium, 0.5% magnesium, and 0.2% iron. As a rule of thumb, all of these elements must be available in the environment for cellular growth and reproduction to occur.

While individual bacteria cells are microscopic, colonies composed of many cells may be easily seen. A common procedure for counting individual cells in water involves spreading a measured volume of water on a plate of agar (a nutrient-containing growth media). Wherever a viable cell adheres to the plate, a colony forms. These visible colonies are counted and related to the number of cells initially present.

7.2.2   Fungi

Fungi are nonphotosynthetic organisms which usually have a filamentous structure. Fungi are chemo-organotrophic, deriving their energy from organic chemical reactions. Some fungi are single cells such as microscopic yeasts (unicellular), while others develop into large forms like "toad stools." Generally, the microscopic fungi (5–10 microns) are much larger than bacteria.

Fungi are aerobic (require oxygen) and usually thrive in more acidic media than bacteria. They also are usually more tolerant than bacteria to elevated concentrations of heavy metals. Classification of fungi is based mainly on the structure of the hyphae (branches), formation of sexual spores, production of fruiting bodies, development cycle, and chemical nature of the cell walls.

Fungi are most useful in the environment as destructors of cellulose in wood and other plant materials. Fungi release a biological catalyst (similar to enzymes) called cellulase that converts cellulose into soluble carbohydrates which can be absorbed by the fungal cell.

Fungi grow better in soil than water and cause large amounts of organic degradation. A primary end product produced by fungi is humic material, which interacts with cationic ions and hydrogen.

7.3 AUTOTROPHIC AND HETEROTROPHIC ORGANISMS

Autotrophic organisms do not depend on organic matter for growth and live well in inorganic media. They extract carbon for their growth from carbon dioxide or other carbonate species. Energy for their metabolic processes may be derived from a variety of sources, depending on the bacteria species. A biologically mitigated reaction, however, is always responsible for the energy.

An example of autotrophic bacteria is *Gallionella*. In an aerobic setting, these bacteria grow in the presence of oxygen; they also can thrive in a medium consisting of ammonium chloride, phosphates, mineral salts, carbon dioxide (carbon source), and solid iron sulfide (energy source). The following reaction is the energy yielding reaction:

$$4FeS(s) + 9O_2 + 10H_2O \rightarrow 4Fe(OH)_3(s) + 4S_4O^{-2} + 8H^+$$

Autotrophic bacteria must synthesize all of the complicated proteins, enzymes, and other materials needed for life processes. While the biochemistry of autotrophic bacteria is very complicated, their consumption and production are responsible for many geochemical transformations.

Heterotrophic bacteria depend upon organic compounds as energy sources. Heterotrophs are much more common than autotrophs. These bacteria are responsible for breakdown of organic matter. Fungi are entirely heterotrophic, deriving carbon and energy by degradation of organic matter. Algae are autotrophic organisms, which use carbon dioxide as a carbon source and light as an energy source.

7.4 BACTERIA MEDIATED OXIDATION AND REDUCTION REACTIONS

Bacteria may be categorized based on terminal electron acceptors. Energy generation for chemotropic bacteria requires the oxidation of a substrate.

Oxidation is defined as the reduction of electron state by the addition of oxygen or removal of electrons. For organotrophic bacteria, energy is obtained through the oxidation of an organic substrate. For lithotrophic bacteria, energy can be created by oxidizing an inorganic substrate. For example, the bacteria *Thiobacillus* oxidizes sulfur to form sulfuric acid and, in the process, creates 236 Kcal of energy/mole reactant according to the following reaction:

$$2S + O_2 + 2H_2O \rightarrow 2H_2SO_4$$

To maintain thermodynamic mass balance, for every oxidation there must be a corresponding reduction. Therefore, for oxidation to occur there must be a compound capable of receiving the transferred electrons. These so-called electron acceptor compounds can include oxygen, sulfate, Fe^3, phosphate, nitrate, CO_2, and certain organics.

Aerobic bacteria are those which reduce oxygen by the addition of protons (H^+ ions) to form H_2O. Bacteria capable of metabolism in the absence of oxygen are *anaerobic*. Some microbiologists refer to methanogenic bacteria (methane production through the oxidation of organics or CO_2) only as anaerobic. In this classification scheme other bacteria, such as iron (III) or sulfate reducers, are designated as *anoxic* types. Where this is applied, the term anoxic refers to the utilization of oxygen by these bacteria from oxidized compounds like goethite (FeOOH) or sulfate (SO_4^{-2}).

Some bacteria do not fit neatly into either aerobic or anaerobic types. *Facultative* bacteria are principally anaerobic types that can tolerate oxygen. Microaerophilic bacteria are bacteria that are principally anaerobic (anoxic), but need small quantities of oxygen for metabolism.

All direct bacterial oxidation and reduction reaction couples are enzymatically mitigated. Enzymes serve as a catalyst which lowers the threshold energy for oxidation and permits reactions to occur at moderate temperatures. The transfer of electrons is not a simple one-step process. Rather, it is accomplished through a complex set of intermediate processes collectively known as an electron transport system.

Energy produced through an oxidation/reduction reaction is stored in very complex organic molecules such as adenosine triphosphate (ATP), guanosine triphosphate (GTP), and acetyl-coenzyme A. By far the most important of these is ATP. For this reaction, adenosine diphosphate (ADP) acquires an additional phosphate group to form ATP. This type of reaction is called phosphorylation, and requires energy input. This energy is stored in the phosphate bond, and can be released by breaking the bond at a later time.

7.5 KINETICS OF GROWTH

The rate of growth of a bacterial colony is directly related to the available nutrients, carbon source, and energy source. Figure 7.3 describes a typical microbial growth curve for a setting where the nutrient supply is limited (closed environment).

Initially, the cells require a period of time to adjust to their environment. No multiplication occurs until the cells have nearly doubled in size. After this initial period of adjustment, the cells begin to reproduce exponentially. Every time a cell divides, a new generation is created, doubling the population. Some cells can reproduce every 10 minutes, some every 20, while others require a full day.

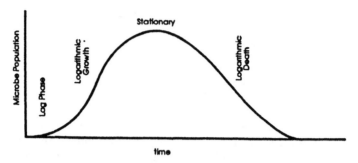

Figure 7.3   Characteristic bell-shaped bacterial growth curve.

In addition to the hereditary background, the availability of nutrients, water, and proper temperature are also factors controlling the reproductive rate.

The rapid growth phase is called the logarithmic growth phase. During this phenomenal growth phase, a single cell with a generation time of 20 minutes could produce 2.2 x 10^{43} cells per day. Approximately one trillion cells weigh one gram. If this rate of reproduction continued for one full day, the mass produced would weigh more than the weight of the earth. Obviously, this cannot occur; the growth phase is controlled by the food supply and production of waste products. As the growth slows and eventually stops, the stationary phase is reached, where no net reproductive growth occurs.

If food supplies are not renewed, or wastes removed, the colony will enter the logarithmic death phase. The reduction in numbers will be at a rate similar to the rapid growth on the earlier side of the curve. Finally, the entire culture dies and the cycle is complete.

The discussion above is applicable to a closed system. Bacterial growth in an open system, where substrate, nutrients, and electron acceptors can be resupplied, is typically described using the Monod equation:

$$\mu = \frac{\mu_{max} * S}{K_s + S}$$

where:

μ = Specific bacterial growth at a point in time
μ_{max} = Maximum bacterial growth
S   = Available substrate at a point in time
K_s = Half velocity constant (the concentration of substrate at $0.5\mu_{max}$.

Figure 7.4 shows typical bacterial growth relative to the Monod equation. As the concentration of substrate increases, the rate of microbial growth increases, approaching the limit presented by μ_{max}. Once μ_{max} is achieved, the addition of more substrate does not result in a corresponding increase in specific

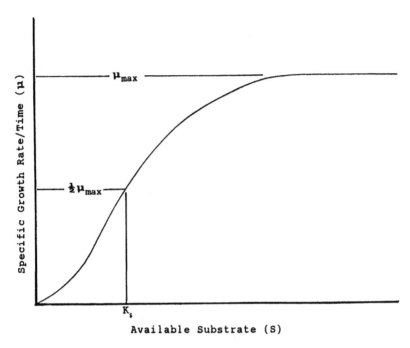

Figure 7.4   Bacterial growth in an open system.

bacterial growth. As $\mu \rightarrow \mu_{max}$, other limitations, such availability of space or electron acceptor transfer rate, serve to limit population growth.

Temperature can affect bacterial growth kinetics. Different varieties of bacteria function best within a set temperature range:

$$Psychrophilic = < 20° \ C$$
$$Mesophilic \quad = 20° - 40° \ C$$
$$Thermophilic = > 45° \ C$$

All bacteria operate within certain cardinal temperatures including a minimum, optimum, and maximum temperature. As a rule of thumb, the level of bacterial growth and corresponding metabolic rates will double for each 10°C above the minimum temperature threshold value.

7.6   MICROBIAL TRANSFORMATION OF CARBON

The interaction of microbes with various carbon compounds is a part of the "carbon cycle" shown graphically in Figure 7.5. Removal of carbon from the atmosphere is the result of utilization of carbon dioxide released by the weathering of minerals or by photosynthesis (where carbon dioxide is absorbed by plants and oxygen is liberated). Carbon dioxide is returned to the atmosphere by the decay of organic matter, combustion, and animal respiration. The organic decay process results in the liberation of carbon and hydrogen in the forms of carbon dioxide, methane, and water (etc.).

Large quantities of "new" carbon dioxide are naturally released by volcanos and mineral springs. Since the industrial revolution, the carbon dioxide concentration in the atmosphere has increased steadily due to use of fossil fuels and reduction of vegetative areas which reprocess the carbon dioxide.

Carbon is a basic building block of life. For most organisms, the bulk of net energy-yielding or energy-consuming metabolic forces involves changes in the oxidation state of carbon.

Chemical transformation of carbon is important; for example, when algae or other plants fix carbon dioxide as a carbohydrate (symbolized as $\{CH_2O\}$), carbon changes from the +4 to 0 oxidation state.

$$CO_2 + H_2O \rightarrow \{CH_2O\} + O_2$$

When algae die, bacterial decomposition results in the reverse reaction; energy is released and oxygen is consumed.

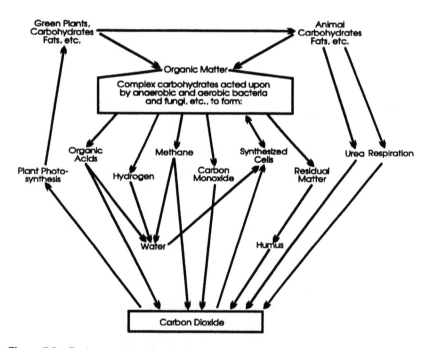

Figure 7.5 Environmental carbon cycle.

In aerobic settings, the main energy-yielding reaction of bacteria is oxidation of organic matter, expressed as:

$$O_2 + \{CH_2O\} \rightarrow CO_2 + H_2O$$

The free energy change of this reaction is 30 KCal per electron mole. This type of reaction provides bacteria and other microbes with the energy necessary for their growth and reproduction.

Decomposition of organic matter usually proceeds in a stepwise manner, which is a major process in the formation of peat, lignite, coal, oil shale, and petroleum. Under reducing conditions (anaerobic), the oxygen content of the original organic matter is decreased, leaving relatively higher carbon content.

7.7   METHANE-FORMING BACTERIA

In anoxic sediments where low concentrations of other alternative electron acceptor compounds are present, methane production is favored. Methane production is a key role in the final step of anaerobic decomposition of organic matter. Almost 80% of the methane in the atmosphere was produced by anaerobic digestion.

Carbon in methane can be derived from carbon dioxide or fermentation of organic matter. When carbon dioxide is the electron acceptor in the absence of oxygen, the general chemical equation is:

$$CO_2 + H^+ + e^- \rightarrow CH_4 + H_2O \quad \text{(unbalanced)}$$

This reaction is mediated by methane-forming bacteria. When organic matter is degraded under anaerobic conditions,

$$\{CH_2O\} + H_2O \rightarrow CO_2 + H^+ + e^- \quad \text{(unbalanced)}$$

When the above two reactions are added together chemically, the overall end result describes the degradation process by methane-producing bacteria:

$$\{CH_2O\} \rightarrow CH_4 + CO_2 \quad \text{(unbalanced)}$$

In real cases, the above reaction is a series of complex step reactions. Fermentation is the redox process where both the oxidizing agent and reducing agent are organic substances. Fermentation produces only about 20% as much free energy as is obtained by the complete oxidation of a similar amount of organic matter by aerobic bacteria.

Methane-producing bacteria are "obligately anaerobic" (by necessity without molecular oxygen). Four main categories are: Methanobacterium, Methanobacillus, Methanococcus, and Methanosarcina. Methane formation is a valuable process for degradation of large quantities of organic matter in nature. These bacteria degrade organic matter at locations where free oxygen is not available, such as in the subsurface environment.

7.8   BACTERIAL UTILIZATION OF HYDROCARBONS

Methane is oxidized under aerobic conditions by a number of varieties of bacteria. *Methanomonas* is a highly specialized strain that cannot utilize any other carbon source for its energy needs. Methanol, formaldehyde and formic acid are intermediate chemical steps toward microbial oxidation of methane to carbon dioxide.

$$CH_4 \rightarrow CH_3OH \rightarrow CH_2{=}O \rightarrow OHCH{=}O \rightarrow CO_2$$

Several varieties of bacteria are able to degrade higher hydrocarbons for use as carbon and energy sources. Hydrocarbon degrading bacteria are widespread in all natural environments, about 1–10% of natural species have been found to be capable of degrading hydrocarbons. These bacteria include *Micrococcus, Pseudomonas, Mycobacteria,* and *Myocardia.* Degradation of hydrocarbon by these organisms is an important environmental process for removal of petroleum contaminants in water and soil.

The most common step in the oxidation of alkanes (straight chain hydrocarbon structure with single bonds) is the conversion of a terminal R-CH$_3$ to a R-C=OOH (acid) group. Some rare initial enzyme attacks involve addition of an oxygen atom to a non-terminal carbon thus forming a ketone (R-C=O-R).

After formation of a carboxylic acid (R-C=O-OH), further oxidation normally occurs by a process illustrated by the following reaction.

$$CH_3CH_2CH_2CH_2CH_2C{=}O{\text -}OH + 3O_2 \rightarrow CH_3CH_2CH_2C{=}O{\text -}OH + 2CO_2 + 2H_2O$$

Following a series of complex reactions, the residue at the end of each cycle is an organic acid with two fewer carbons than at the beginning of the cycle.

Degradation potential of hydrocarbon molecules varies. Microbes have a strong preference for straight chain compounds. One reason for this preference is that branching inhibits oxidation at the site of a branch. Structures such as the following are difficult to degrade.

$$
\begin{array}{c}
CH_3 \\
| \\
R{-}C{-}CH_3 \\
| \\
CH_3
\end{array}
$$

Aromatic rings (benzene rings) are rather easy to degrade. The overall stepwise process of ring breaking is:

Actual breakage of the ring is preceded by addition of -OH to adjacent carbon atoms. *Cunninghamella elegans* is a microscopic fungi which attacks aromatic rings. It metabolizes a wide range of hydrocarbons including C_3 to C_{32} alkanes (straight chain compounds), alkenes (double bond compounds), and aromatic compounds (ring compounds, i.e., benzene, toluene, and naphthalene).

Biodegradation of aromatic compounds is also important when related to polynuclear aromatic hydrocarbons (PNAs) such as benze(a)pyrene and other toxic varieties.

The physical form of spilled hydrocarbons makes a great difference in its degradability. Degradation occurs at the water-oil interface with bacterial enzymes and oxygen. For example, thick layers of crude oil prevent contact of the ingredients necessary for efficient interaction. In response to this phenomena, bacteria often produce an enzyme that causes the oil to be dispersed as a fine colloid, and thus accessible to the bacteria cells. Sometimes the enzymes act as a surfactant to release free phase oil which was previously attached to soil particles.

7.9 NITROGEN TRANSFORMATIONS BY BACTERIA

Nitrogen is an essential building block of proteins, including enzymes. While it is the most abundant component of air (78% by volume), it cannot be directly used by higher plants or animals. Nitrogen is a rather unique element in that it can exist at valences of -3 to $+5$. For example, from the zero state, as N_2, it can lose as many as five electrons and exist as nitrate (NO_3^-), or it can gain three electrons and exist in its reduced state as ammonia (NH_3) or ammonium (NH_4^+). Many bacteria are able to capitalize on nitrate's unique properties.

Some of the most important bacterially facilitated reactions in the soil environment are those involving nitrogen compounds. These reactions include N-fixation where N is fixed as organic N; nitrification where ammonia (NH_3) is oxidized to nitrate (NO_3^-); nitrate reduction where NO_3^- is reduced to form compounds

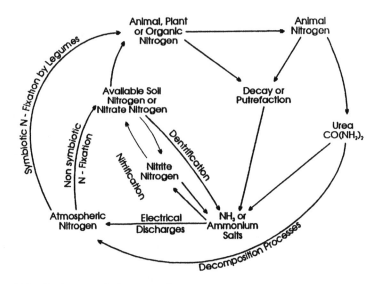

Figure 7.6 Environmental nitrogen cycle.

which have N in a lower oxidation state; and denitrification (reduction of NO_3 and NO_2 to N), which releases N_2 to the atmosphere.

Constant withdrawal of nitrogen from the soil by plants would soon deplete the essential supply if it were not replenished by microbes. The nitrogen cycle is the dynamic process through which N is recycled between the atmosphere, organic matter, and inorganic matter. Figure 7.6 is a diagrammatic presentation of the N cycle.

7.9.1   Nitrogen Fixation

The biological process for fixing nitrogen from the atmosphere to the chemically bound form is:

$$3\{CH_2O\} + 2N_2 + 3H_2O + 4H^+ \rightarrow 3CO_2 + 4NH_4$$

The actual process steps used by the microbes to complete this reaction is *very* complicated and not completely understood. Biofixation of N is an essential process for plant growth in the absence of synthetic fertilizers.

Very few microbial species have the ability to fix atmospheric N. The best known and most important N-fixing microbe is *Rhizobium*, which has a symbiotic relationship with leguminous plants such as clover and alfalfa. The bacteria are found in nodules attached to the root structure of the plant. The microbes derive energy from the plant and in turn deliver usable fixed nitrogen to the plant. When the legumes die and decay, NH_4^+ is released back into the soil. The NH_4^+

is in turn converted by other microbes into the nitrate ion which is usable by other plants.

Some nonlegume angiosperms fix N with the aid of actinomycete bacteria contained in root nodules. Several species of trees and shrubs are hosts to these bacteria.

Free-living bacteria (i.e., *Spirillum lipoferum*) associated with some grasses are stimulated by the grasses to fix nitrogen. In tropical settings, the quantity of nitrogen fixed by these microbes can amount to nearly 100 kg per hectare per year.

7.9.2   Nitrification

Aquatic nitrogen in thermodynamic equilibrium with the atmosphere is in the +5 oxidation state (NO_3^-), while in most biological systems, nitrogen is present as N (-3), as in $-NH_2$ in amino acids. The general overall reaction is:

$$2\ O_2 + NH_4^+ \rightarrow NO_3^- + 2\ H^+ + H_2O$$

Nitrification (conversion from ammonia N to nitrate) is especially important because nitrogen is absorbed by plants as nitrate. When agricultural fertilizers are added to soil, microbes transform them to nitrate for easy assimilation by plants.

In nature, nitrification is catalyzed by two groups of bacteria: Nitrosomonas and Nitrobacter. *Nitrosomonas* cause conversion of ammonia to nitrite.

$$NH_3 + 3/2\ O_2 \rightarrow H^+ + NO_2^- + H_2O$$

Nitrobacter facilitate the oxidation of nitrite to nitrate:

$$NO_2^- + 1/2\ O_2 \rightarrow NO_3^-$$

Both of the nitrogen bacteria are strictly aerobic. Each can also utilize oxidizable inorganic materials as electron donors in oxidation reactions to yield needed energy for metabolic processes.

7.9.3   Nitrate Reduction

When N is changed from the +5 oxidation state to the -3 oxidation state, the process is called reduction. In the absence of free oxygen, some bacteria are able to use NO_3^- as an electron acceptor. Nitrogen is an essential element of protein. Any organism which uses nitrogen from nitrate for synthesis of protein must first reduce the N to the -3 oxidation state.

Generally, however, when nitrate ions function as an electron receptor, the end product is nitrite (NO_2^-):

$$2 \ NO_3^- + \{CH_2O\} \rightarrow NO_2^- + H_2O + CO_2$$

7.9.4   Denitrification

Denitrification is a special case of nitrate reduction in which the final product is nitrogen gas (N_2). This reaction proceeds as (per electron mole):

$$1/5 \ NO_2^- + 1/4 \ \{CH_2O\} + 1/5 \ H^+ \rightarrow 1/10 \ N_2 + 1/4 \ CO_2 + 7/20 \ H_2O$$

Denitrification is an important process in nature, in that it is the return of free nitrogen to the atmosphere. Because N gas is nontoxic and volatile, and since the NO_3^- ion is a very efficient electron acceptor, denitrification encourages extensive bacterial growth under anaerobic conditions.

7.9.5   Competitive Oxidation of Organic Matter by Nitrate Ions and Other Oxidizing Agents

Successive oxidation of organic matter by dissolved O_2, NO_3^-, and SO_4^{-2} causes varied responses of nitrate ion levels in sediments, especially when oxygen was initially available but not replenished when exhausted.

While oxygen is present, some nitrate may be generated. After free molecular oxygen is exhausted, nitrate is the favored oxidizing agent. When the nitrate has been consumed, sulfate becomes the primary electron receptor. Actually, these steps are not perfectly distinct (as efficient mixing is not common in soils); however, the oxidation of organic matter will continue as long as possible.

7.10 BIOTRANSFORMATION OF SULFUR

Sulfur-containing compounds are very common in soil and groundwater. The sulfate ion (SO_4^{-2}) is found in varying concentrations in almost all natural water. Organic sulfur compounds (both natural and pollutants) and degradation of these compounds is an important microbial task. In some circumstances, degradation products (such as H_2S) are odiferous or toxic.

A strong parallel exists between the sulfur and nitrogen cycles in nature. In living organisms, sulfur is present in its reduced state as -SH (the hydrosulfide group). Nitrogen in living matter is in the −3 oxidation state. When decomposed by bacteria, the initial sulfur product is generally the reduced form, H_2S. When nitrogen compounds are reduced by microbes, the reduced form is NH_4^+.

In the same way that microbes produce elemental nitrogen, some bacteria produce and store elemental sulfur from sulfur compounds. In the presence of molecular oxygen, some bacteria convert reduced forms of sulfur to the oxidized state (SO_4^{-2}), while other bacteria catalyze the reduced nitrogen to nitrate ion.

The sulfur cycle is a complex mixture of microbially mediated processes, geochemical processes, and SO_2 in air pollution.

7.10.1 Oxidation of H₂S and Reduction of Sulfate by Bacteria

While organic sulfur compounds are often the source of H_2S in groundwater, they are not the only source. *Desulforibrio* bacteria can reduce sulfate ions into H_2S. The sulfate ion can be an electron acceptor. The general overall reaction is:

$$SO_{4-2} + 2\{CH_2O\} + 2H^+ \rightarrow H_2S + 2\ CO_2 + H_2O$$

In actuality, other bacteria must assist *Desulforibrio* with the ultimate conversion of organic matter into carbon dioxide. The oxidation of organic matter by *Desulforibrio* usually ends with acetic acid. Other bacteria complete the process. In sediments where sulfide formation occurs, the sediment is often dark in color because amorphous pyrite (FeS_2) is formed abiotically.

In gypsum and limestone formations, the biologically mediated reduction of calcium sulfate deposits produces elemental sulfur which is often then dispersed in the pores of the rock. The general equation is:

$$2\ CaSO_4 + 3\ \{CH_2O\} \rightarrow 2\ CaCO_3 + 2\ S + CO_2 + 3\ H_2O$$

The calculated quantity of $S(s)$ is never found because H_2S formed as an intermediate product often escapes into the atmosphere.

While some microbes can reduce H_2S to elemental S, others can oxidize H_2S to higher oxidation states. Purple sulfur bacteria and green sulfur bacteria derive energy for their metabolism from oxidation of H_2S. These bacteria use CO_2 for their carbon source, and are strictly anaerobic.

Aerobic colorless bacteria use molecular oxygen to oxidize H_2S:

$$2\ H_2S + O_2 \rightarrow S + 2\ H_2O$$

$$2\ S + 2\ H_2O + 3\ O_2 \rightarrow 4\ H^+ + 2\ SO_4^{-2}\ \text{ or}$$

$$S_2O_3^{-2} + H_2O + 2\ O_2 \rightarrow 2H^+ + 2\ SO_4^{-2}$$

Oxidation of sulfur in a low oxidation state to a (higher) sulfate ion produces sulfuric acid. *Thiobacillus thiooxidans*, a colorless sulfur bacteria, can tolerate acid solutions up to 1N in strength. When elemental S is added to alkaline soils, bacteria mediate transformation into sulfuric acid.

7.10.2 Microbial Degradation of Organic Sulfur Compounds

Sulfur is a component of many types of biological compounds. The degradation of these compounds is an important part of the natural recycle program.

Sulfur-containing functional groups found in groundwater include: hydrosulfide (-SH), disulfide (-S-S-), sulfide (-S-), sulfoxide (-S-), sulfonic acid (-SO$_2$OH), thioketone (-C-), and thiazole (a heterocyclic sulfur group). Protein contains some amino acids with sulfur functional groups. These amino acids are readily broken down by bacteria and fungi.

Biodegradation of sulfur-containing compounds (such as amino acids) can produce volatile organo-sulfur compounds such as methyl thiol (CH$_3$SH) and dimethyl disulfide (CH$_3$-S-S-CH$_3$). These compounds have a very unpleasant odor. In addition to H$_2$S (rotten egg gas), these chemicals account for the odor associated with biodegradation of sulfur-containing organic compounds.

7.11 IRON BACTERIA

Several types of bacteria, including *Ferrobacillus*, *Gallionella*, and some forms of *Sphaerotillus* use iron compounds to obtain metabolic energy. These bacteria catalyze the oxidation of Fe^{2+} to Fe^{3+} by molecular oxygen:

$$4\ Fe^{2+} + 4H^+ + O_2 \rightarrow 4\ Fe^{3+} + 2\ H_2O$$

The carbon source for some of these bacteria is carbon dioxide. Since they do not require organic matter or carbon, these bacteria may thrive in areas where organic matter is absent.

Microorganism mediated oxidation of iron Fe^{2+} is not an efficient means of obtaining energy for metabolic processes:

$$FeCO_3 + 1/4\ O_2 + 3/2\ H_2O \rightarrow 3/2\ Fe(OH)_3 + CO_2$$

The release of free energy in this reaction is approximately 10 KCal/electron mole. Approximately 220 grams of Fe^{2+} must be oxidized to produce 1 gram of carbon. The production of 1 gram of cell carbon produces almost 430 grams of solid Fe(OH)$_3$. Large volumes of hydrated Fe^{3+} oxide form in areas where iron bacteria thrive.

Some bacteria (*Gallionella*) secrete large volumes of hydrated Fe^{3+} oxide in the form of branched structures. The bacteria cells grow at the end of the twisted stalk of iron oxide.

At near neutral pH, bacteria deriving their energy from the aerobic oxidation of Fe^{2+} must compete with direct chemical oxidation of Fe^{2+} by O$_2$. As a consequence, the iron bacteria grow in a thin layer between the oxygen source and the iron source.

The similarity of chemistry between iron and manganese results in similar reactions between bacteria and these elements. While the overall scale of activity is less with manganese, it is not uncommon to find manganese bio-oxidation present near that of iron bacteria processes.

7.12   MICROBIAL DEGRADATION OF PESTICIDES

Biodegradation of pesticides in the environment is quite important to preservation of environmental quality. Herbicides and insecticides generally have little effect on microbes. However, effective fungicides must be antimicrobial to function well. Often, fungicides harm beneficial fungi which decompose dead matter (saprophytic fungi) and bacteria.

Biodegradation of pesticides (and similar compounds) by microbes occurs in complex, stepwise, microbially catalyzed reactions. These steps include oxidation, reduction, hydrolysis, and dehalogenation (removal of Cl and Br atoms) most often mediated by microbes.

The large number of different chemical pesticides and the wide variety of degrading microbes are beyond the scope of this text. The key concept is that degradation can and does occur. The reader is referred to advanced texts and papers for details.

## CHAPTER 8

Sampling Techniques

8.1 INTRODUCTION

Collection of soil samples for investigation or monitoring involves both science and art. The science part requires that the investigator include sufficient quality control procedures in the project plan to assure that variables associated with sampling are identified and quantified. In theory, after analysis of the samples has been completed, any variability of data represents the actual distribution of parameters in the soil. The skill of the art is to recognize that the qualities of subsurface material are not always definable in conventional mathematical terms. Also, in most cases the project budget is not adequate to provide the number and type of samples required to assure the least risk spread of data results.

Use of statistical procedures to aid interpretation of resulting numerical data can determine the accuracy of the data plus the confidence limits. Identification and usage of recommendations for the wide variety of statistical procedures currently available is beyond the scope of this text. Readers who wish to pursue the study of statistical analysis should consult standard texts and EPA publications. Several good references are included in the bibliography.

8.2 SAMPLING CONSIDERATIONS

An acceptable sampling program must include collection procedures which produce high quality samples, representative of the variability of site conditions, within the project budget. The project plan should include a detailed discussion of the objectives of sampling, and uses of the resulting data.

A soil sampling effort may be conducted for a variety of reasons, some which include:

* Characterization of background site conditions
* Determination of the distribution of specified parameters across a defined area
* Compliance with cleanup criteria (for remediation projects)
* Data collection as part of a research or environmental model validation study, and/or
* Identification of the source of a contaminant release, mode of transport, or possible receptors

Selection of the "best" sampling procedure is often a complex task based on the goals and objectives of the sampling plan. Guidance for preparation of a sampling can often be obtained by reference to previously successful similar projects. Consultation with experienced environmental consultants and regulatory agencies is the most important first step in plan preparation.

8.3   SAMPLING STRATEGIES

8.3.1   Site Investigation

The main purpose of a general site investigation is to define the variability of soil across the site and to compare the results against the "normal" condition of the site (i.e., the benchmark against which all environmental measurements are compared). In most instances, background determination is a component of the overall site investigation. Selection of the number and type of soil samples collected is determined by the general objectives of the study, the parameters of interest, general geology, land use, and the degree of confidence required.

In many cases, such as a prepurchase environmental audit, the initial records search may indicate that contamination is highly unlikely. Under this scenario, a minimum number of samples (three or four) collected at locations of most likely contamination may be adequate. At an industrial site known to be contaminated, the site investigation program should be adequate to identify the contaminant, define "hot spots," determine the boundary of the contamination, and provide additional data to evaluate potential health risks.

It is prudent, therefore, for the investigator to prepare a formal sampling plan which meets all of the objectives of the project before initiating the field work. Several general types of sampling strategies are discussed in the following paragraphs.

Grab Samples

This type of sample is often collected by the investigator to develop a "feel" for the site. Most often, selection of sample locations is made by professional judgment at either interesting or unique spots. The number of samples, or the type of analysis to which they are subjected, also is determined by the professional.

When collected by an experienced investigator, grab samples can be an effective technique. The key factor is the ability of the professional to recognize and evaluate probable site conditions. Judgments of this variety are not mathematically defensible and should be reserved for special situations. Grab samples are the most difficult to defend during litigation.

Pattern Sampling

Another technique for site evaluation is the preselection of sampling locations at discrete locations. A common practice is to construct a grid pattern on a site map and select sample locations at node points (see Figure 8.1). This procedure is particularly effective at sites which are relatively open (no obstructions) and the potential for contamination (or site variability) is unknown. The reliability of grid sampling depends on the number and spacing of the grid pattern. If a ten-acre site is sampled at the surface on two-foot centers, the likelihood of missing a significant surface chemical spill is very small. On the other hand, if the same site is sampled on a 800-foot grid, only large surface spills are likely be discovered.

A variation of grid sampling is to construct the grid patterns on the map, then select a predetermined number of nodes at random as sample locations. The number of samples can be varied, based on the degree of confidence required and the project budget.

Likely Location Sampling

Probably the most widely used procedure to select sample locations at previously developed sites is to make a two (or three) step site selection effort.

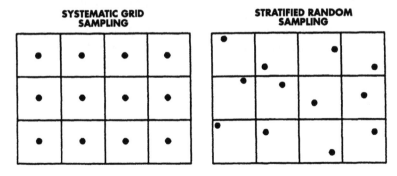

SYSTEMATIC GRID SAMPLING   **STRATIFIED RANDOM SAMPLING**

Figure 8.1 Grid sampling.

The initial step is to review all available site data prior to identifying any discrete sample sites. If possible, a thorough site walkover inspection is recommended. Data developed during the initial review can be reduced to identify the most likely locations for representative sample collection. This procedure allows the investigator to collect samples which represent the overall character of the site, both contaminated and background. Follow-up tasks may be necessary to complete the process, but the selection of the initial locations is easily defensible.

Composite or Individual Samples

Much discussion has been devoted to the value of composite samples. By definition, a composite sample is prepared by blending two (or more) samples and removing an aliquot which represents the larger sample. When this procedure is used, the number of analyses may be reduced.

Composite samples have the advantage of averaging the distribution of data. If one is interested in the overall character of a site, the use of composite samples is justified. A common practice in the agricultural soil sampling is to select a number (10–15) of discrete soil samples from a field, blend them into a single large composite sample, and extract an aliquot for nutrient analysis. This type of sampling provides reasonably representative depiction of the nutrient concentration in the field of interest. The farmer may have confidence that the quantity of fertilizer determined by this type of analysis will be adequate for acceptable crop yields.

On the other hand, blending of samples tends to disguise both high and low concentration values. If nine average concentration samples are blended with one high (or low) value sample, the nontypical sample loses its significance.

In the practice of environmental science, the outlying sample value is often the most sought result. If the project plan does not allow for an adequate number of samples to identify areas of concern, composite samples may still be used to increase coverage. A technique often employed is to consider the minimum detection limit expected for the laboratory and the minimum acceptable concentration for health consideration. The established health risk concentration divided by the analytical concentration limit defines the maximum number of samples which can safely be composited while still assuring adequate detection.

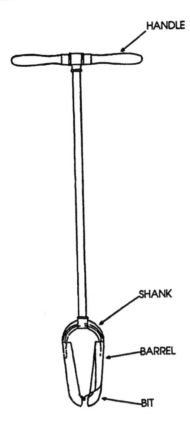

HANDLE

SHANK

BARREL

BIT

Figure 8.2   Posthole type of barrel auger.

8.4   SAMPLING TECHNIQUES

8.4.1   Shoveling and Drilling

The sampling utensil selected to collect a soil sample depends on the purpose of the sample, depth of sample, volume of soil required, and the type of soil. When a grab sample is collected from surface soil for general analytical purposes, it is common to use a shovel or similar implement. Shovels excavate a small quantity of bulk soil. This variety of sample is "disturbed" (that is, not maintained in its in-place form); however, for general sampling, it is an appropriate procedure.

Other manual procedures, such as posthole augers or small diameter hand augers, are often used to collect deeper disturbed samples where the soil is easily excavated and is noncollapsing. Figure 8.2 shows a typical posthole type auger. Extensions to the shaft can be added in increments to allow for deeper penetration. Most manual sampling is limited to depths of less than 20 feet; at greater depths, the physical labor output required for sampling increases so much that power equipment is usually used.

Power augers are the most common subsurface sampling units. Solid flight augers allow for continuous removal of cuttings from the boring. These augers come in a wide variety of forms. Some are constructed with solid cores which have attached helical flights (see Figure 8.3). Augers are usually manufactured in common lengths of 3, 5, or 10 feet. Solid stem sampling augers are commonly available in diameters of 3, 4, 6, and 8 inches.

When solid stem augers are used for sampling, they are twisted into the soil to the desired depth. Grab samples may be collected from the cuttings as they arrive at the surface; however, the exact depth from which the soil was drilled cannot be assured. An alternate procedure is to twist the auger to the desired depth, then withdraw the auger (without twisting) and observe the soil retained to the augers. If the soil is sufficiently cohesive so that it clings to the auger, it is possible to make some observation of the vertical soil distribution. Again, the sample is disturbed and suitable only for general purposes. Variations in texture, layering, or other structure are not preserved. In addition, if the soil is noncohesive, or not capable of supporting boring walls, the hole may collapse and limit further sampling.

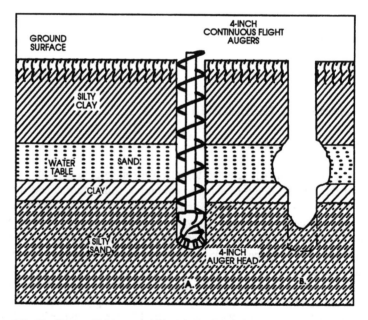

Figure 8.3   Continuous flight auger drilling through coring material.

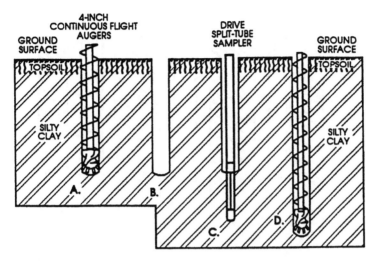

Figure 8.4   Continuous flight auger drilling.

The most satisfactory procedure for sampling with solid stem augers is to advance the auger to the sample depth, then retrieve the augers and insert a drive or push-type core sampling tube (see Figure 8.4). When the core tube is pushed below the bottom of the borehole and retrieved, the exact location of the sample is known. Also, when the core is extracted from the tube, the sample may be seen in its original orientation.

Difficult sampling conditions include boreholes which are drilled into wet sand or in noncohesive soils below the water table. When the borehole is not self-supporting, alternative sampling methods are recommended.

A variation of auger sampling which compensates for caving soils is the hollow stem auger which has the flights welded to the outside of a central hollow tube. On this type of auger (see Figure 8.5), cutting teeth are attached to the perimeter of the hollow stem. A separate center rod, with attached teeth, is inserted inside the hollow stem tube. As the auger is advanced, cuttings are conveyed up the outside by the flights. When the desired depth is achieved, the center rod is removed, leaving the augers in place to support the borehole walls. A core sampler can be inserted to the total depth, pushed or driven into virgin soil below, and retrieved with a soil sample.

The hollow stem auger sampling procedure is a rapid and effective method of sample collection. Standard procedures are described in ASTM D 1452. Hollow stem augers are commonly available with inside diameters ranging from 2 1/2 to 6 inches, and outside diameters up to 12 inches.

A relatively recent improvement to the hollow stem auger sampling system is the continuous core unit (see Figure 8.6). This unit is similar to a standard

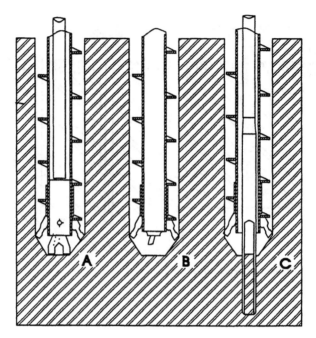

Figure 8.5 Drilling and sampling with hollow stem augers.

hollow stem auger, except that the center core tube is held in place and advanced with the augers. The core barrel is the same length as the lead auger section (usually 5 feet) and does not rotate with the augers. Because it extends slightly ahead of the auger teeth, it is able to retrieve a relatively undisturbed sample.

After the continuous core unit is advanced the length of a full flight, the upper part of the unit is disassembled and the core barrel is retrieved. Because the sample is the same length as an auger flight, the progress of a boring project is not retarded by unnecessary tool-removing trips.

Other equipment occasionally used to collect soil samples includes the standard mud rotary rig, air rotary rig, and sometimes cable tool drilling rigs. These procedures are all adaptations of water well drilling equipment, and each uses some form of fluid at the bottom of the boring to conduct the soil sample to the surface. Under normal circumstances, if any other procedure is available, these techniques are not recommended for environmental soil sample collection.

8.4.2   Undisturbed Samples

Collection of core samples during soil boring procedures may be accomplished by the use of several standard sampling tools. Each of these tools was developed to retrieve the best "undisturbed" sample for specific purposes.

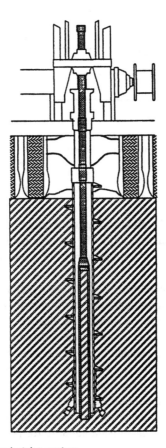

Figure 8.6 Continuous sample tube system.

The traditional split spoon sample tube is constructed of heavy steel wall tube which is split from end to end so that it may be opened to reveal a core sample. The upper end is equipped with a threaded connection which is attached to standard drill rod connectors. The lower end has a steel drive shoe to protect the sampler. A standard split spoon sampler is specified by ASTM D 1586. With an inside diameter of 1 3/8 inches and an outside diameter of 2 inches, the sampler can withstand severe usage. These units are manufactured in lengths of 18, 24, and 36 inches. Other split spoons are available in larger diameters and longer lengths.

Split spoon samplers may be pushed or driven into the soil. A standard driving hammer for split spoons weighs 140 pounds. When the hammer is dropped 30 inches and the number of blows required to cause 6 or 12 inches penetration is recorded, the resulting data are an indication of the shear strength of the soil. Blow counts, then, provide additional information which may be useful to the

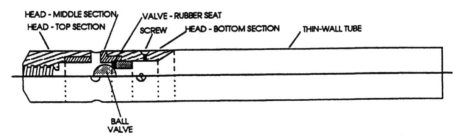

Figure 8.7   Thin-walled (Shelby tube) sampler.

investigator, such as the ability of an excavation or boring to stand open without collapse, to indicate the difficulty of excavation with conventional equipment, or the approximate bearing strength of a soil for structure foundation design.

Soil samples removed from a split spoon may be used for chemical analysis, moisture content determination, *in situ* density calculation, or particle size analysis. Special liner tubes (brass and plastic) may be installed in the split spoon barrel. When a liner is used, the soil sample is contained in the liner. After the spoon is retrieved and opened, the liner may be removed quickly and capped for shipment to the laboratory. As the sample has minimal exposure to the atmosphere, there is little opportunity for environmental changes to occur. Fewer organic volatile components are lost by evaporation, and exposure to temperature changes or introduction of microbes can be controlled. Also, on occasion, these samples have been suitable for laboratory vertical permeability tests.

Another core sampling method is the use of a thin wall sampler which is often called a "Shelby Tube." This sample tube is manufactured from thin steel tubing which is adapted for attachment to standard drill rods (see Figure 8.7). The cutting edge of the sampler is reduced in size to contain the cut soil sample. Typically the cutting throat has a reduced area of from 0.5 to 1.5%. Details of construction are described in ASTM 1587.

When a thin wall sampler is used, the tube is attached to a sufficient length of drill rod to reach the bottom of the bore hole. Once in place, the tube is *slowly* pushed to the desired sample length and then retrieved. At the surface, the sample may be capped or sealed with warm beeswax for shipment to the laboratory. The main advantage of thin wall tubes compared to split spoon samples is that the thin wall construction causes much less disturbance of the sample during collection. The primary disadvantage of this type of sample collection is that it is most suitable for use in cohesive soils (clay). Sand samples often drop from the tube during sample retrieval.

Thin wall sample tubes are produced in diameters of 2, 3, and 5 inches. Lengths are readily available in 24 and 30 inches.

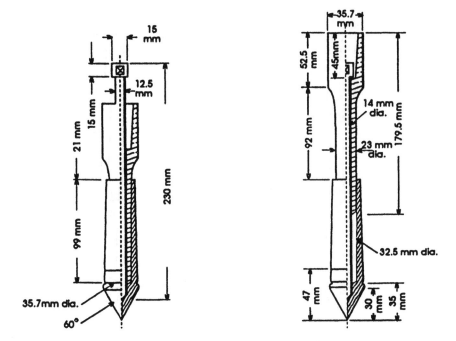

Figure 8.8 Example of a mechanical cone penetrometer tip (Dutch mantle cone).

8.4.3 Probes

Some types of information about the subsurface do not require the collection of actual soil samples, but can be measured by the use of probes. Soil shear strength and relative density can be determined by use of a cone penetrometer (sometimes referred to as a Dutch cone). A standard tool used in this method is described in Figure 8.8. This procedure involves slowly pushing an expendable probe point into the soil, and monitoring the effort required. Charts of effort versus depth and time versus depth can be evaluated to indicate the location of stratification changes and shear strength. At least *some* actual borings are recommended to confirm the findings and provide an opportunity for calibration of the results.

A variation of penetrometer sampling involves the attachment of plastic tubing to the expendable perforated probe point (see Figure 8.9). The tubing extends through hollow push rods to the surface. After the probe has been inserted to the desired sampling depth, the push rods are withdrawn a measured distance, exposing the portholes in the sides of the probe point. With the direct connection to the surface, soil vapor or water samples (above suction lift depth) can be retrieved.

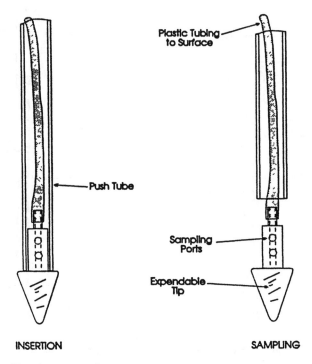

Plastic Tubing to Surface

Push Tube

Sampling Ports

Expendable Tip

INSERTION   SAMPLING

Figure 8.9   Water/vapor sampling penetrometer.

A major advantage of probe sampling procedures is that they are very rapid. Truck-mounted probe equipment is often able to complete many more tests per day than conventional soil auger sample procedures to similar depth.

8.5   VOLUME OF SAMPLES REQUIRED

Each laboratory procedure requires a specific quantity of soil to assure that the analytical procedures can be completed accurately. Table 8.1 presents the quantity of sample required for several commonly prescribed physical testing procedures. This table was prepared from published data and conversions; however, it is recommended that the investigator communicate with the laboratory during the project planning stage to assure that adequate sample volumes are collected. The quantity of sample required for chemical testing is discussed in Chapter 9. It is much less expensive to dispose of excess soil samples than to return to the field for additional soil, or to explain in the report why the analytical results do not meet the desired testing standards.

Table 8.1 Recommended Minimum Sample Sizes

1. Particle Size Analysis:

Largest Particle Diameter		Minimum Size of Sample	
(ins)	(mm)	(gm)	(lb)
3/8	9.5	500	1
3/4	19.0	1000	2
1	25.4	2000	4
1 1/2	38.1	3000	6
2	50.8	4000	8
3	76.2	5000	10

2. Compaction Density Test: 50–100 lb. (23–45 kg)

3. Atterberg Limits: 2 lb. (1 kg)

4. Moisture Content:

Particle Size		Sample Size	
(mm)	(sieve)	(gm)	(lb)
2.0	10	300	0.6
4.75	4	500	1.0
19.0	3/4	1000	2.0

5. Classification of Soils by USCS:

Maximum Particle Size		Sample Size	
(mm)	(sieve)	(gm)	(lb)
4.75	4	100	0.25
9.5	3/8"	200	0.5
19.0	3/4"	1000	2.2
38.1	1 1/2"	8000	18.0
75.0	3"	60,000	132.0

## CHAPTER 9

Selection of Analytical Procedures

9.1 INTRODUCTION

The investigation of soils at a site may require any of a wide variety of testing procedures, depending on the overall objectives of the project. Field observations are usually insufficient to define the necessary parameters, therefore requiring tests to be made under controlled laboratory conditions.

Laboratory testing may include physical testing to define partial size distribution, moisture content, stress-strain relationships, or a host of other mechanical properties. Other analyses may be performed to quantify trace concentrations of hazardous organic chemicals or measure the exchangeable ion capability. Over the years, analytical procedures have been developed which provide the best results for all of the major analyses commonly required by environmental professionals.

The primary purpose of analysis is to provide meaningful, accurate data. Representative sampling procedures were discussed in Chapter 8. In the same way, results of laboratory testing should accurately represent the sample provided, and the results should be expressed in a form which can be interpreted by the end user.

This chapter is organized into sections which discuss testing procedures, equipment, and quality control. The common denominator of all varieties of analysis is that of quality control; thus, this topic will be discussed first.

9.2 QUALITY CONTROL-QUALITY ASSURANCE (QA-QC)

The two terms, quality control and quality assurance, are often confused and used inappropriately. Quality control (QC) is a set of measures which assures that the data meets "control" limitations. Procedures of quality control often include making duplicate analysis, running blind (unknown content) samples, inserting spikes (known quantities of analyte added to the sample), comparison with known standards, and other tests to confirm that analytical results are accurate and precise.

Quality assurance (QA) is a management system for ensuring that all information, data, and decisions resulting from the project are technically sound and properly documented. QA, then, is essentially an audit function to ensure that the QC work has been satisfactorily completed.

The process of establishing criteria for a specific QA/QC program begins with the definition of the decisions to be made based on the data and the overall objectives of the project. The second step is to define a set of criteria that then can be used to design the analytical plan and to establish the appropriate level of quality assurance. Ultimately, the goal of the QA/QC program is to determine the reliability of the data. This process should be initiated early in the project, *before* any significant investment is made in analysis.

Most companies and government agencies involved in environmental soils work have prepared or adapted a detailed QA/QC program to ensure that their scientific work meets acceptable standards. These plans include detailed analytical procedures, as well as specifications for the number of duplicate/replicate samples, specification of instrument calibration standards, and all of the other required monitoring parameters. The essential elements of a QA/QC program are presented in Table 9.1. Fortunately, most of the actual procedures have been standardized and published by government and industry agencies.

Several key terms are used to describe the range of analytical results. A brief description follows:

Precision: a measure of the scatter of a group of measurements made at the specified conditions about their average, or in other words, measurement of the degree of agreement among analysis of replicate analyses of a sample. Values calculated should demonstrate the reproducibility of the measurement process. The sample standard deviation and sample coefficient of variation are commonly

Table 9.1. Essential Elements of a QA/QC Plan

1. Detailed Project Description
2. Project Organization and Responsibility Definition
3. Quality Assurance Objectives
4. Sampling Procedures
5. Sample Custody Procedures
6. Instrument Calibration Procedures and Frequency
7. Analytical Procedures
8. Data Reduction, Validation and Reporting
9. Internal Quality Checks
10. Performance and Systems Audits
11. Maintenance Program
12. Specific Routine Procedures Used to Assess Data
13. Standards of Precision, Accuracy and Completeness
14. Corrective Actions for Failed Data
15. Quality Assurance Reports to Management

used as indices of precision. The smaller the standard deviation and coefficient of variation, the better the precision.

Precision is stated in the units of measurement or as a percentage of measurement average, as a plus and minus spread around the average measured value. Sources of variation or error may result from sample collection, handling, shipping, storage, preparation, and analysis.

Accuracy: a statement of the bias that exists in a measurement system or, the agreement of a measurement against an accepted reference or true value. Accuracy is normally expressed as the difference between measured and reference (or true values), or the difference as a percentage of the reference value. It may be expressed as a ratio of the measurement to the true value.

The determination of accuracy (or bias) within a measurement system is generally accomplished through the analysis of the blank sample and the analysis of the sample which has been spiked with a known concentration of a standard reference material.

The point at which the sample is spiked determines which component of the measurement system has an effect on the accuracy of the analysis. The three most common spiking points are: in the field (acquisition spike); during laboratory preparation (lab matrix spike); and analysis (analysis matrix spike). The field matrix spike provides the best estimate of bias based on recovery. It includes matrix effects associated with sample preservation, shipping, preparation, and analysis. The lab matrix spike provides an estimate of recovery incorporating matrix effects associated with sample preparation and analysis only. The analysis matrix spike provides an indication of matrix effects associated with the analysis process only.

Bias: usually expressed as "measurement bias," is a statement of consistent under- or overestimation of the true values in population units. If the measurement of each unit is consistently 2 ppm too high, then a positive measurement bias of 2 ppm is present.

Figure 9.1 is a graphic presentation of the relationships between bias, precision, and accuracy.

Completeness: the measure of the amount of valid data obtained from a measurement system compared to the amount which was expected to be obtained under normal conditions. Completeness goals should be defined at the beginning of the project to ensure that sufficient valid data are received.

Representativeness: the degree to which data accurately and precisely represent a characteristic of a population, parameter variations at a sampling point, or an environmental condition. All testing data should reflect as best possible the conditions which existed at the time of sample collection.

Comparability: an expression of the confidence with which one data set can be compared with another. The produced data should be compared to other available data such as: data generated by an outside laboratory over a specific time period; or data collected from a totally independent source (literature or research by others).

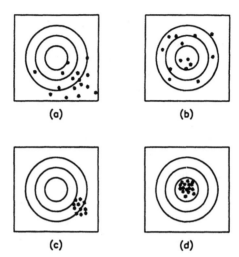

Figure 9.1   Patterns of shots at a target (after Jensen, 1978, Figure 1). (a): high bias + low precision = low accuracy; (b): low bias + low precision = low accuracy; (c): high bias + high precision = low accuracy; (d): low bias + high precision = high accuracy.

Whenever measurement data are evaluated, the reviewer should keep the levels of concern and the end use of the data in mind. In some cases, even data of poor precision and/or accuracy may be useful. For example, if all the results are far above a level of concern, the precision and accuracy are much less important. However, results which are close to the level of concern, precision, and accuracy are much more important and should be closely reviewed.

9.3   PURPOSE OF THE SOIL TESTING

The initial task of any environmental investigation is to define the purpose of the project. General types of projects include basic soil characterization, health risk evaluation, remediation-restoration, agricultural field testing and research. Each type of project has its own specific data needs. The degree of precision and accuracy required will vary according to the data quality objectives.

9.4   MECHANICAL TESTING

When soil testing is required for particle size distribution, permeability, or other mechanical properties, several sources of standard procedures are available. Civil engineers have long recognized the need for standardization of testing methods, and thus all of the common testing procedures are very similar.

The most common compendiums of civil engineering soil testing procedures are the American Society for Testing and Materials, and the U.S. Bureau of Reclamation. Other sources of specific tests include the U.S. Navy engineers' manual and other industry-specific manuals.

Table 9.2 is a listing of common soil testing procedures included in Section 4 of the Annual Book of ASTM Standards. Procedures in this volume are updated on a regular basis.

9.5 CHEMICAL TESTING

Any discussion of chemical analysis of soils or soil components must be divided into categories for understanding. When the decision is made to submit a

Table 9.2 ASTM Standards Referenced in This Text

D 1452	Practice for Soil Investigation and Sampling by Auger Borings
D 1586	Method for Penetration Test and Split-Barrel Sampling of Soils
D 1587	Practice of Thin-Walled Tube Sampling of Soils
D 2573	Test Method for Field Vane Shear Test in Cohesive Soil
D 3441	Method for Deep, Quasi-Static, Cone and Friction-Cone Penetration Tests of Soil
D 4220	Practices for Preserving and Transporting Soil Samples
D 5195	Test Method for Density of Soil and Rock In-Place at Depths Below the Surface by Nuclear Methods
D 421	Practice for Dry Preparation of Soil Samples for Particle-Size Analysis and Determination of Soil Constants
D 422	Method for Particle-Size Analysis of Soils
D 3404	Guide to Measuring Matric Potential in the Vadose Zone Using Tensiometers
D 4700	Guide for Soil Sampling from the Vadose Zone
D 5126	Guide for Comparison of Field Methods for Determining Hydraulic Conductivity in the Vadose Zone
D 698	Test Method Laboratory Compaction Characteristics of Soil Using Standard Effort
D 854	Test Method for Specific Gravity of Soils
D 1140	Test Method for Amount of Material in Soils Finer than the No. 200 (75-μm) Sieve
D 1557	Test Method for Laboratory Compaction of Soil Using Modified Effort
D 4318	Test Method for Liquid Limit, Plastic Limit and Plasticity Index of Soils
D 2434	Test Method for Permeability of Granular Soils (Constant Head)
D 3152	Test Method for Capillary-Moisture Relationships for Fine-Textured Soils by Pressure-Membrane Apparatus
D 5084	Test Method for Measurement of Hydraulic Conductivity of Saturated Porous Materials Using a Flexible Wall Permeameter
D 5093	Test Method for Field Measurement of Infiltration Rate Using a Double-Ring Infiltrometer with a Sealed Inner Ring
D 2166	Test Method for Unconfined Strength of Cohesive Soil
D 4972	Test Method for pH of Soils
D 2487	Test Method for Classification of Soils for Engineering Purposes
D 2488	Practice for Description and Identification of Soils (Visual-Manual Procedure)
D 2922	Test Methods for Density of Soil and Soil-Aggregate in Place by Nuclear Methods (Shallow Depth)

soil sample for chemical procedures, the first basic decision must be what general property is being sought.

Some chemicals are adsorbed to the surface of the soil grains. Extraction of these loosely held chemicals can be accomplished by use of a suitable solvent. Once the analyte is in solution, the solution may be analyzed for the compound(s) of interest. On the other hand, to determine the total lead content of a native soil, it is necessary to destruct or destroy the molecular complexes of the entire soil sample (by dissolving the sample with acid). Again, the solution is analyzed.

The intermediate variety of analysis is the most difficult. If the purpose is to determine the presence of an exchangeable cation, the extraction procedure must be sufficiently severe to remove the exchangeable cation, without extracting any of the similar cations from the soil matrix. Often, this balance can be a very difficult task. Significant experience is required to derive meaningful data from this type of analysis.

9.6 GENERAL PROCEDURES (ORGANIC CHEMICALS)

Once the desired analyte is in solution, several alternatives are available for analysis of almost any compound. Most procedures are most accurate within a limited range of concentrations. If the concentration of analyte is expected to be in the percent range, the analytical procedure selected will be different than that for trace concentration.

9.6.1 Gas Chromatography

The presence of organic chemicals may be quantified by any of several methods. The most common procedure is that of gas chromatography, which is completed by a gas chromatograph (GC). This procedure is based on the principle that when a mixture of volatile materials is transported by a carrier gas through a column containing an absorbent material, each volatile component will be partitioned between the carrier gas and the absorbent. The length of time required for a volatile compound to travel the length of the column is proportional to the degree to which it is retained by the absorbent.

Because different analytes are retained by different degrees, they have different travel times (called retention time). A detector located at the end of the column records the time of arrival and quantity of component. A graphical printout of the detector data appears as peaks of different heights at different times on the graph. The area enclosed by the peaks is proportionate to concentration.

When a known compound at a predetermined concentration is analyzed to calibrate the GC, it is possible to make very accurate concentration determinations. Figure 9.2 is a representative chromatograph.

Figure 9.3 is a schematic diagram of a gas chromatograph. The carrier gas is usually argon, helium, or nitrogen. Precise flow control of the gas flow rate is essential for accurate analysis.

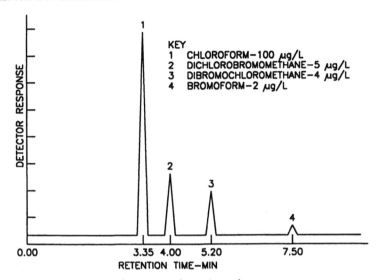

Figure 9.2 Sample readout from a gas chromatograph.

9.6.2 Purge and Trap

A variation of standard GC instrumentation is "Purge and Trap." This procedure is used for volatile compounds which have low boiling points and are easily vaporized. Solutions containing these chemicals are placed in a small airtight (or airless) vessel, and the carrier gas is allowed to bubble up through the liquid to purge the organics and transport them onto a trapping column. The trapping column contains an absorbent which collects the organic compounds. The trapping column is heated to volatilize the collected components, which are then passed into the GC.

Most GC procedures require that several calibration compounds be used; thus, it may be possible to evaluate several compounds during one run. A restriction to this practice is that the separate compounds must arrive at the detector at different times with definite separation between component peaks.

9.6.3 Detectors

The type of detector attached to the end of the GC is selected by the variety of chemicals being analyzed. A photoionization detector (PID) works by stimulating the gasified organic molecule to emit a light spectrum, then measuring the variety and intensity of light waves produced. The pattern of emissions is distinct for each separate compound. PIDs are designed specifically for aromatic compounds (ring structures). Other detectors often used for organic compounds are the flame ion detector (FID), thermal conductivity detector (TCD), the Hall detector, and electron capture detector (ECD).

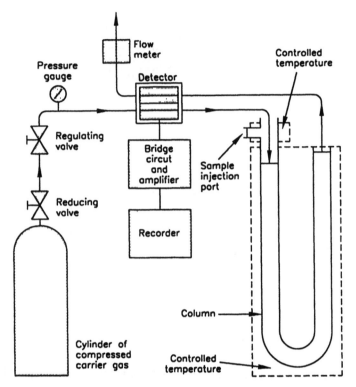

Figure 9.3 Schematic diagram of a gas chromatograph with thermal conductivity detector. Large arrows indicate direction of gas flow.

9.6.4   Gas Chromatograph Mass Spectrometer

In many situations the content of the solution is not known. A very good screening tool is the Gas Chromatograph-Mass Spectrometer (GCMS). A GCMS is a combination of GC, which is used to separate the various compounds, and a mass spectrometer (MS), which is used in this situation as the detector. An MS measures the atomic mass of the components which enter from the GC. The pattern of retention times and atomic mass is distinct for each compound. GCMSs are connected to a computer which has a "library" of compound-specific patterns in storage. The pattern produced by the instrument is compared to the library, which allows identification of the compounds in solution, as well as the concentration.

GCMS procedures are widely used for initial screening of the soil's organic compounds. It is not uncommon to determine the presence of 50 or more compounds. After the compounds present have been identified, successive testing is usually made for specific analytes by GC methods. GC techniques tend to have better quantitative capabilities and are significantly less expensive.

9.7   GENERAL PROCEDURES (INORGANIC)

9.7.1   Titrimetric (Wet Chemistry)

Historically, determination of the elemental content of soil has been an important process for understanding the "mysterious ways" of the natural world. Initially, chemical procedures were completed by "wet" procedures, i.e., the dissolving of elements, separation by precipitation and resolution, and weighing the various resulting reaction products. This type of analysis is very accurate to low parts per million ranges; however, it requires an experienced chemist, is time-consuming, and expensive. Few commercial laboratories continue to provide this service.

9.7.2   Spectrophotometry

Modern practice is based on the principles of output, i.e., spectral absorption and transmittance. Several common types of instruments are available with differing degrees of accuracy and detection limits.

Reasonably accurate concentration data can be achieved by the use of a visible spectrophotometer. This procedure involves the solution of the chemical of interest, and mixing with specific reagents. A beam of light is passed through the reacted sample and then through a prism. Characteristics of the emitted spectrum can be compared to a known standard which determines the presence and concentration of the chemical. Figure 9.4 is a schematic diagram of a single-beam spectrophotometer.

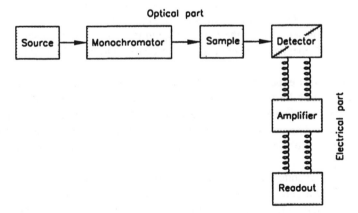

Figure 9.4   Block diagram showing components of a single-beam spectrophotometer. Arrows represent radiant energy, coiled lines electrical connections. The optical part and the electrical part of the instrument meet at the detector, a transducer which converts radiant energy into electrical energy.

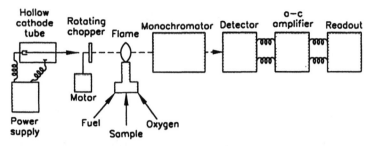

Figure 9.5 Components of an atomic absorption spectrophotometer. (The flame may be replaced with a furnace.)

9.7.3   Atomic Absorption (AA) Spectrophotometry

Two general types of AA analytical procedures are commonly used for metals analysis. Both types are based on the absorption of monochromatic light by a cloud of vaporized atoms of the analyte metal.

In the flame AA procedure (Figure 9.5), a beam of monochromatic light produced by a hollow cathode ray tube passes through a flame charged with the sample. Metal atoms in the flame intercept some of the radiation from the cathode tube beam as the excited electrons in the sample atoms jump to higher orbitals. Variations of the beam can be measured by analyzing variations of the energy beam, either as absorption or emission of light energy (due to changes in electron excitement). After passing through a system of optical devices, the beam enters a detector which converts the light energy into an electrical signal which is representative of the metal concentration in the original sample. Flame AA is rapid, inexpensive, and accurate at concentration in the parts per million range.

Difficulties experienced with the injection of a liquid sample into a flame (i.e., control of the flame, and uniform distribution of the sample throughout the flame) are the primary reasons for the relatively high detection limits of the Flame AA method. The use of a graphite furnace as nonflame vaporization method reduces these problems and enables quantification of metals in the parts per billion range.

In the Graphite Furnace Atomic Absorption (GFAA) procedure, the flame is replaced by a pure graphite furnace. When the sample is injected into this uniform heat source it is vaporized. Monochromatic light from the hollow cathode tube is passed through the vaporized cloud in the furnace, then processed by the optical and detection instruments using the same theory as absorption mode flame AA.

9.7.4   Atomic Emission Techniques using Inductively Coupled Plasma (ICP)

ICP is a relatively new (since about 1976) variation of emission spectroscopy. The "flame" or heater in this procedure is an incandescent plasma (ionized gas) of argon heated inductively by radio frequency energy. The energy is transferred to a stream of argon through an induction coil, which produces temperatures up to 10,000°K. The sample atoms are heated to temperatures around 7,000°K (about twice as hot as an oxy-acetylene torch). Light emissions from the superheated atoms are passed through an optic system to separate the spectrum, and detectors record the response. As many as 30 elements can be analyzed simultaneously.

9.7.5   Electrochemical Analysis

Electrochemical sensors, also called ion selective electrodes (probes), are often used to determine the specific ion content of a sample solution. These probes may be designed on the principles of potentiometry or voltametry (i.e., how they conduct electrical current). The general principle is the relationship between the electrical potential of a measuring electrode and that of a reference electrode. If the components of an electrode are properly selected, concentration measurements can be made with reasonable accuracy over a fairly wide range of concentrations (parts per million to percent). Electrode sensors are available for several ions; among them are ammonia, bromide, calcium, chloride, cyanide, fluoride, iodide, potassium, and sodium.

9.8   ANALYTICAL PROCEDURES

Methods of analysis have been developed for almost every conceivable purpose. Some procedures, very serviceable for general data generation, produce results which are accurate within percent ranges. This type of data is very useful for bulk characterizations where the purpose of testing is to define the proportions of the most significant components. If the analysis must determine the concentration of a particular analyte, alternative procedures will be required. Other factors which must be considered include the possibility that the analyte may be adsorbed on the surface of the soil grains or only loosely attached by chemical attraction.

Analytical procedures have been fairly well standardized for most common purposes. Analyses of water and waste (both solid and liquid) are presented in the EPA SW-846 Procedures Manual. Table 9.3 lists the procedures prescribed by EPA SW-846, along with the minimum sample sizes required.

Projects which are subject to regulatory review or are expected to be scrutinized by litigation should be performed under either of the EPA-prescribed procedures. These methods have been accepted as standard practice.

Table 9.3. Methods of Chemical Analysis

Parameter	EPA SW-846 Series	Sample Volume	Sample Container	Notes
Chloride	9250	10 g	4 oz	
Cyanide	9210	10 g	4 oz	
pH	9040/9045	100 g	4 oz	
Cr (VI)	7196	10 g	8 oz	
Mercury	7470	10 g	8 oz	
Metals [except Cr(VI) and HG]	7000 Series or 6010	10 g	8 oz	
Nitrate	9200	10 g	8 oz	
Nitrate/Nitrite	8200	10 g	8 oz	
Oil/Grease/TPH	9070	10 g	8 oz	
Organic Carbon	9060	10 g	4 oz	
Phenol	9066	10 g	8 oz	
Specific Conductance	9050	100 g	8 oz	
Sulfate	9035–38	10 g	4 oz	
Sulfide	9030	100 g	8 oz	
Purgeable halocarbon	8010/8240	5 g	4 oz	Cool to 4°C
Purgeable aromatic hydrocarbons	8020	5 g	4 oz	Pack tightly in container, Cool to 4°C
Acrolein and acrylonitrile	8030	5 g	4 oz	
Phenols	8040	10 g	8 oz	
Benzidines	8270	10 g	8 oz	
Phthalate esters	8060	10 g	8 oz	
Nitrosamines	8270	10 g	8 oz	
PCBs	8080	10 g	8 oz	
Nitroaromatics and isophorone	8090	10 g	8 oz	
Polynuclear aromatic hydrocarbons	8100	10 g	8 oz	
Haloethers	8270	10 g	8 oz	
Chlorinated hydrocarbons	8120	10 g	8 oz	
Dioxins and furans	8280/8290	10 g	8 oz	
Total organic halogens (TOX)	9020	10 g	4 oz	
Pesticides, chlorinated	8080/8140	10 g	8 oz	
Gross Alpha	9310/9315	100 g	8 oz (HDPE)	
Beta	9310	100 g	9 oz (HDPE)	
Nonhalogenated volatile organics	8015/8240	5 g	4 oz	Pack tightly in container, Cool to 4°C
Semivolatile organics	8250/8270	10 g	16 oz	Cool to 4°C
TCLP	1311	100 g	16 oz	
Hazardous waste characteristics	1010/9010 1110/9030	200 g	8 oz	

Notes: 1. Glass containers are preferred, with Teflon-lined lids, except where indicated.
        2. HDPE indicates High Density Polyethylene container.

Agricultural and research projects are often more result-oriented and less subject to review by nontechnical persons. Selection of methods for these special projects may involve minor variations in procedure which enable the analyst to produce results in a specific format or focus on a specific aspect of analysis. The best compendium of general soil analysis is "Methods of Soil Analysis, Part 2 - Chemical and Microbiological Properties." This extensive volume contains a practical procedure for analysis of almost any need of the environmental scientist.

Selection of the analytical procedures for a specific project should be made as part of the overall project plan. Considerations in the selection should include:

* Definition of the actual data needs of the project: if a general screening is all that is required, the detection limits need not be to parts-per-billion level.
* Consultation with a knowledgeable chemist (not necessarily the analyst) or microbiologist who understands the purpose of the analysis.
* Definition of the quality control/quality assurance plan; it is important to determine that the results will pass the scrutiny of the project reviewer (or client).
* A decision as to whether later phases of the project will require data that could be easily developed by use of another procedure, at little or no additional cost.
* Careful review of the most promising methods and documentation of reasons for the method selected.

CHAPTER **10**

Agricultural Considerations

10.1 INTRODUCTION

In previous chapters, the study of soils has focused on specific aspects of soils based on individual characteristics across the entire soil column. This chapter is devoted to that upper layer of soil which is responsible for the growth of plants and the preservation of life on earth as we know it. The soil and life evolved together. There can be no soil without life and no life without soil.

Historically, the study of soils has been largely centered on agricultural production. Soil science has been a key factor in the understanding of how the soil functions. The agricultural properties of soil rely upon all of the individual scientific properties working in harmony to maintain the equilibriums that sustain the various "life cycles."

While all of the major scientific disciplines are important, many are discussed in significant detail in previous chapters. The main topics presented in this chapter are extensions of the most important, which are focused on the needs of agriculture.

10.2 ORGANIC MATTER

Most garden variety soil contains between 1 and 5% organic matter by weight. Exceptions to this statement are numerous, especially in areas where peat soils are common. Peat soils, however, comprise a relatively small percentage of the overall land surface.

Organic matter in soil is derived entirely from plants and animals that once lived in this soil. Most of the organic matter results from decomposition of green plants, while a substantial quantity results from the recycling of microscopic organisms. As one variety of organism dies and is reduced to simpler compounds, other organisms utilize the resulting material as building blocks for their own construction.

175

Under conditions of adequate aeration, the degradation of organic matter tends to be more complete. Peat soils, which are often composed of greater than 90% organic matter, were deposited in very wet areas or ponds where oxygen content was minimal. Under these anaerobic conditions, the breakdown of the organic matter is minimal, often resulting in preservation of some of the leaf and limb structures. Organic matter tends to increase with lower average soil temperature, which retards degradation processes. A clay soil in a warm climate will typically contain less organic matter than a similar soil in a cooler climate. Similar conditions occur in sand or other soil.

Light-colored sandy soil usually contains around 1.0 to 1.5% organic matter, while light-colored silt and clay loam may contain around 2%. Dark-colored silt and loam typically contain 4 to 10% organic matter.

As previously noted, organic matter has a low specific weight but a very high surface area per unit weight. A very small quantity of organic matter can have a very large effect on the water-holding capacity, supply of available nutrients, and provide grain binding structure which resists erosion.

One of the primary values of organic matter is that when microbes partially degrade it, releasing its vital by-products, it causes the soil to be a "living host," to have "structure" that permits easy water entry and pore spaces which are connected to the atmosphere. Soils which contain low concentrations of organic matter are often poor crop lands. The reintroduction of organic matter through cultivation or mulch cover can significantly improve crop production.

10.3 FACTORS AFFECTING PLANT GROWTH

Agriculture is a highly technical science devoted to the production of food, fiber, and intermediate chemical products (raw materials for the chemical industry). Soil is an integral part of this production, as it is the source of many of the raw products, and provides the structural foundation necessary to support the operating units (plants).

Essentially, raw materials provided by the soil (fertilizers, crop residues, manure, etc.) are converted by the plants into complex chemical compounds. Several factors are especially important for the efficient growth of plants:

Moisture (water supply)
Energy from a light source
Air—both above and below ground
Temperature
A supply of nutrients
Soil reaction (acidity or alkalinity)
Control of insects and predators
Control of diseases
Genetic factors

A brief summary of the importance of each of these factors is presented below.

10.3.1 Water

While up to 70 to 90% by weight of green plants is water, this is a small percentage of the total quantity needed for crop production. Typically, one pound of dry alfalfa may consume approximately 700 pounds of water; one pound of dry corn requires about 350 pounds of water.

In the process of plant growth, water is required for formation of sugars and carbohydrates; this synthesis occurs as hydrogen from the water is combined with carbon dioxide. Water is also taken in through the roots and passed upward through the plant structure and out through the leaves. This passage of water from the plants results in "transpiration," which causes evaporative cooling of the leaves. The transpirational stream also conveys nutritive materials from the soil into the plant. The inner tissues of the plant are thus supplied with water containing the essential nutrients. There is also a downward migration of some of the water which transfers sugars and other manufactured products to other parts of the plant to promote growth and for storage.

The type of plant which will grow in any specific location is dependent upon the specialized organs which are matched to the environmental conditions of the site. Rice and cranberries are adapted to grow well in very wet areas, while cacti are designed for areas of low water availability. Good land management considers the factors which will favor the movement and storage of water in the soil suitable for the growth of the desired plants.

10.3.2 Light

Photosynthesis is the chemical reaction of carbon dioxide and water which is energized by light to produce sugars and plant tissue. All green plants require light for growth. It is estimated that corn utilizes only about 1.6% of the available sunshine for actual growth. Approximately 44% of the radiant energy is used in transpiration of water.

10.3.3 Air

The presence of air is also important for plant growth. With a standard natural composition of 78% nitrogen, 21% oxygen, and 0.3% carbon dioxide, air often contains a small percentage of sulfur dioxide. Carbon dioxide, sulfur, and nitrogen are the most important airborne plant nutrients (in addition to water).

The composition of soil air is also very important. Roots of plants and microbes in the soil also require oxygen. If the soil is compacted, contains sufficient clay or other low permeability material, or is waterlogged, it will not be possible for oxygen to penetrate the soil. Plant growth in poorly aerated soils is restricted to specialized species, usually not the desired varieties.

While man usually cannot regulate the quality of air on a very localized area, the quality of soil air may be enhanced by proper land management (such as proper tillage or prevention of compaction). Careful management of soil air is essential to favorable crop production.

10.3.4 Temperature

Certain temperatures are required for seed germination and plant growth. For most agricultural crops, the soil temperature must be above 45°F for growth. Biological activity both above and below ground is speeded up with higher temperature. Soil is an effective heat storage unit which provides a moderating effect for roots and subsurface microbes. Vegetative growth continues continuously day and night.

The rate of heat absorption during the day is dependent upon the color of the soil, length of daylight, angle of sunlight incidence to the soil, plant cover over the soil, and moisture content. It is indeed fortunate that spring sunshine in northern climates often encounters relatively bare soil which allows maximum heat absorption. Extreme soil temperatures in summer are limited where plant cover is extensive or mulch (of some variety) is added.

10.3.5 Plant Food

When the term food is used for human or animal consumption, it refers to the major intake of caloric producing materials such as starch, protein, or carbohydrates. When applied to plants, food is used to define the 16 or so chemical elements which plants require for growth. Each element has a different function within the plant and is required by different plants in varying quantities, depending on the species and the growth phase.

Starting with the basic reaction of carbon dioxide and water in the presence of chlorophyll with the energy provided by the sun to manufacture sugar, plants produce hundreds of different compounds. While the dry weight of plants may be reduced to 95% carbon dioxide, oxygen, and hydrogen, the remaining 5% is composed of the mineral elements which are also required for growth and reproduction. The primary 16 essential elements needed for plant growth are listed in decreasing order of concentration:

1. Carbon: the most abundant (approximately 45% by dry weight)
2. Oxygen: makes up about 43% of green plants by weight
3. Hydrogen: usually constitutes approximately 6%
4. Nitrogen: constitutes around 1-3%
5. Phosphorus: is present in the range of 0.1 to 1.0%
6. Potassium: represents 0.3 to 6.0%
7. Sulfur: is found in concentrations of 0.05 to 1.5%
8. Calcium: makes up 0.1 to 4.0%
9. Magnesium: constitutes 0.05 to 1.5% (10,000 parts per million is equal to 1%)

10. Iron: is found in concentrations of 10 to 1000 ppm
11. Manganese: ranges from 5 to 500 ppm
12. Zinc: represents 5 to 100 ppm
13. Copper: makes up 5 to 50 ppm
14. Boron: is found at 3 to 60 ppm
15. Molybdenum: is present at 0.01 to 10 ppm
16. Chlorine: is typically found in very small concentrations

The first three (carbon, oxygen, and hydrogen) are absolutely necessary for plant life. Nothing happens without adequate availability of these elements.

Nitrogen, phosphorus, and potassium are the elements which are most likely to be present in soil at concentrations which are lower than that required for optimum plant growth. Thus, commercial fertilizer is defined by the concentration of N-P-K.

Sulfur, calcium, and magnesium are considered secondary fertilizer elements because they are often naturally present in soil at adequate concentrations. If lacking, they are easily added along with other fertilizers or additives.

10.3.6 Soil Reaction

Acidity and alkalinity (soil reaction) is an extremely important plant growth factor. One of the most important features of soil reaction is that it often controls the availability of other nutrients. Some elements such as phosphorous are significantly reduced in strongly acid soils, while other elements such as iron and manganese are very high in acid soils. Where the soil is severely alkaline, the availability of iron and manganese is reduced.

10.3.7 Control of Diseases

Growth of plants is often affected by diseases and insects. As these problems are very specialized, they are beyond the scope of this text. Readers who are affected by these problems are referred to their local Agricultural Extension Service agent.

10.3.8 Genetic Factors

This topic is mentioned here because it is also a factor of plant growth. The wide variety of genetic strains developed for each crop type have different nutrient and growth responses. The variety of factors are too wide for this text. Again, the local County Extension Agent is familiar with local conditions and plants which are best suited for the area.

10.3.9 Soil pH

The acidity or alkalinity of soil, as measured by pH, may range from extremes of 4 to 10. Most soils, however, fall within the common ranges of 5.5 to 8.

Both mineral and organic substances contribute to soil reaction. Clay minerals have the capacity of cation exchange, and humic organic fraction holds weak organic acids. Both the mineral and organic fraction give rise to the acidity and alkalinity of the soil, depending on the concentration of replaceable hydrogen or exchangeable base metals (Ca, Mg, Na, and K).

Carbonic acid (HCO_3) is formed by the reaction of carbon dioxide and water. As the slightly acid rainfall percolates through the soil, calcium and magnesium bicarbonates are leached and acid soil is formed as these are replaced by hydrogen. Carbonic acid also reacts with silicate minerals which contain aluminum. Thus, at low pH conditions, aluminum rather than hydrogen would probably be the main exchangeable cation.

When lime is added to an acid soil, some hydrogen and aluminum are replaced by calcium and magnesium. These base exchange reactions are important to prevent calcium, magnesium, and potassium fertilizers from being rapidly leached from the soil. Plant roots exchange hydrogen (through carbonic acid) for the mineral nutrients held by the base exchange materials. This exchange allows plants to absorb the minerals at the rate required. This exchange of base metals, then, acts as a natural buffer system which prevents the soil from becoming very acid or basic over a short time period.

High concentration of organic material in soil increases the number of exchange sites and thus increases buffering capacity. Sands tend to have low buffering capacity, while clay loams and soils high in organic matter (i.e., peat) tend to have greater buffering ability.

Some types of plant material are more acidic than others. Grasses and conifers are noted for their acid forming ability. Under humid conditions, prairie soils are more acid than those formed under hardwood trees. Soils formed under pine trees are more acid than those formed under hardwood trees.

In general, soils formed in areas which have rainfall in excess of 30 inches tend to form acid soils. Where less rainfall is common, the quantity of percolating rainwater is not sufficient to totally displace exchangeable metals, and they accumulate at a shallow depth. The "caliche" layer which forms is alkaline, sometimes resulting in a pH of 8.5 or 9. Treatment of such high pH soil is necessary to improve plant production.

Treatment of alkaline soils usually involves addition of some chemicals such as aluminum sulfate or iron sulfate. When these compounds join with water, sulfuric acid is formed. Another common practice is the addition of elemental sulfur, which is transformed by microbes into acid material.

Acid soils are usually treated by addition of powdered limestone. The reaction of calcium and magnesium carbonate with the acid soil lowers the pH. Treatment of acid soils is much more common than alkaline soils.

Table 10.1 presents a listing of the pH preference of common plants.

Table 10.1   Plant pH Preference

High pH	Moderate pH	Acid pH
Alfalfa	Potatoes	Blue berries
Barley	Red clover	Strawberries
Sugar beets	White clover	Cranberries
Sweet clover	Soy beans	Raspberries
Canning peas	Corn	Azaleas
	Oats, wheat	Camellias
	Beans	
	Blue grass	
	Tobacco	
	Cotton	

10.4 IMPORTANCE OF MAJOR NUTRIENTS (NITROGEN, PHOSPHORUS, AND POTASSIUM)

10.4.1 Nitrogen

If a typical acre of soil at the depth of normal plowing (plow depth) contains 60,000 pounds of organic matter, it probably contains between 500 and 3,000 pounds of nitrogen. The chemical and biological presence of nitrogen in its various forms has been discussed in considerable extent in Chapter 7; these paragraphs explore its application.

Most of the nitrogen uptake by plants is in the form of ammonia or nitrate. For most crops, nitrate is the main usable form.

Nitrogen is an essential part of chlorophyll, the green coloring matter of plants. Each molecule of chlorophyll contains one magnesium and four nitrogen atoms. If the plant suffers from nitrogen deficiency, it will turn light green in color. Nitrogen is also an important component of amino-acids which are part of protein included in plants. Also nitrogen is found in other compounds, such as vitamins.

Nitrogen is the most sensitive plant nutrient other than water. In most crops, a direct relationship exists between the quantity of available nitrogen and the crop yield (within reason). In controlled field tests on corn in the central United States, it was shown that increasing the available nitrogen from 5 pounds per acre to 88 pounds per acre increased the yield from 60 to 110 bushels per acre. Along with the increased yield, the protein increased from 7.1 to 13.1%. Similar experiments with grasses and other grains have demonstrated that the principle is fairly universal.

10.4.2 Phosphorus

Phosphorus is one of the most important elements in plant growth. It is necessary for many of the vital life functions, including the utilization of starch, photosynthesis, nucleus formation, and reproduction. Phosphorus must be present in every living cell. At plant maturity, it is concentrated in seeds and fruits. While the total quantity of phosphorus is rather small compared to other elements, its presence is very significant.

The amount of phosphorus found in soils varies greatly. In some strongly acid sands with low organic content, the phosphorus may only be 80 pounds per acre in the upper 8 inches. At the other extreme, in heavier soils (clays and silts) with a high organic content, the phosphorus content may reach as high as 4000 pounds per acre. Typically, however, the plow layer of most soils may be expected to contain from 800 to 2500 pounds per acre.

The primary characteristic of phosphorus compounds is their general low solubility. Percolating water does not remove great quantities as it migrates through the soil.

Some of the phosphorus in soils is in the form of organic compounds, which were originally formed by plants or microorganisms. The release (or retention) of organic phosphorus is somewhat controlled by the quantity and variety of microorganisms present, and the type of organic matter which is added. For example, if sawdust (low in nitrogen and phosphorus) is added to the soil, the microorganisms attack this food supply. The difficulty is that microbes require both nitrogen and phosphorus; thus, the soil will soon become deficient of these minerals. It is necessary to maintain the proper nutrient balance by the addition of fertilizers from an outside source if dramatic alterations are made in the existing soil system.

Inorganic minerals are present in a wide variety of forms. Native rock phosphate minerals are highly insoluble. They must be altered by the action of mineral acids, soil acids, or organic acids to make them available to plants.

In strongly acid soils, large quantities of iron and aluminum are present. Any phosphate which comes in contact with these minerals reacts to form iron and aluminum compounds. Initially, these newly precipitated compounds are fairly available to plants; however, after a period of time, the precipitates form crystals and become again unavailable. If lime is added, the pH is increased and the phosphorus tends to react to form more soluble phosphate compounds.

The primary conclusion which may be reached from the previous paragraph is that the availability of phosphorus is a complex issue which is dependent upon pH, other minerals, microbial activity, and probably many other factors.

10.4.5 Potassium

One of the primary functions of potassium in plants is to neutralize organic acids. While other base exchange metals (calcium and magnesium) are important soil nutrients, the relative quantity of potassium is fairly large. Potassium has specific duties in plant growth as well; however, no individual organic plant compound has been identified.

Where potassium is deficient, a general buildup of soluble nitrogen compounds occurs and the formation of protein is inhibited. Also, potassium is thought to be essential in carbohydrate metabolism. Within a plant cell, it has definitely been established that potassium is a key chemical in the catalytic reactions of some enzymes. Crops (natural or planted) which are grown in potassium deficient soils exhibit a noticeable reduction of yield, lowering of structural strength, and greater winter kill.

10.5 AVAILABILITY OF WATER FOR PLANT UTILIZATION

After a significant rainfall, all of the soil pores near the surface are filled with water and the water moves downward into the soil, displacing air contained in the pore spaces. At some point, the soil will be saturated so that the soil is at its maximum retention capacity. If the rainfall infiltration continues, the additional water which seeps into the soil will migrate downward, ultimately to the groundwater table. Excess water (greater than the infiltration capacity) will either pond on the surface or flow away as storm water runoff.

After the rain has stopped (the supply halted), the free water (that which is not attached to the soil grains) will continue to migrate downward. After a period of time (hours or days), the relatively rapid downward movement will cease, leaving some water retained in the soil. Some minor downward water migration may occur; however, it is minimal. At this stage of moisture content, the soil is said to be at "field capacity." Water has vacated the larger pores, but the smaller pores are still completely filled.

Growing plant roots can easily extract water from soil at "field capacity." Water also may be lost by evaporation to the soil atmosphere and later dispersion into air above the soil. As the more easily removed water continues to be lost from the soil, the remaining water becomes increasingly more difficult to extract, due to the retentive forces of the soil. Ultimately, the water becomes only slightly available to the plants.

When the water content of the soil reaches the point where the plant root extraction capacity is equal to the soil retentive forces, the soil is said to be at the "wilting point." The moisture content of the soil at the wilting point is not a set value, but varies according to the soil and the plant species growing it. A typical value of the wilting point is a soil suction equivalent to 15 atmospheres.

Water is available to plants only between field capacity and the wilting point. Figure 10.1 shows the water holding properties of several soils in relation to the wilting point and field capacity. The goal of efficient agriculture is to maintain soil moisture between these two values. Soils which contain significant quantities of organic matter, silt, and clay have much higher water holding capacities than sand soils. The most productive soils have a good balance between water holding capacity and sufficient drainage to prevent water logging.

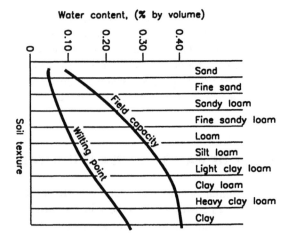

Figure 10.1   Water-holding properties of soils based on texture. The available water supply for a soil is the difference between field capacity and wilting point.

10.6 SOIL CONSERVATION AND EROSION CONTROL

Erosion is the process of transporting soil from its initial location and de-
positing it in another. Some erosion is necessary to facilitate the formation of
new soil. Without erosion, the mineral residue left would be almost incapable of
supporting modern plants. A little slow erosion under virgin plant cover is a valuable process.

Cultivation greatly increases the rate of erosion over that of the native plant
communities. Land which has been "developed" for the benefit of mankind can
be preserved from most erosion by the use of best management practices. Careful planning and management can minimize the rate of erosion and assure that
the land continues to be productive.

The primary agents of erosion of cultivated or bare land are wind and water.
Both of these are most destructive to bare land. Where any type of vegetative
or mulch cover is present, their effect is dramatically reduced.

Wind erosion is greatest where the soil is bare, the surface dry, and the
wind strong. Areas where dry farming is practiced or even in humid regions
where the soil is left uncovered for a period of time are subject to wind action.

The direct force of the wind against soil particles dislodges them from their
resting place. After they have been accelerated by the wind, they tend to spring
forward and upward in a jumping motion. As the particles gain elevation they
also gain momentum, then as they fall, they strike and dislodge other soil particles. As the wind motion is very turbulent near the surface, its action tends to
continue this agitation of soil and air.

Soil movement by the wind results in particle sorting by size. Coarse particles tend to creep along the ground. Smaller size particles are either bounced along or carried in suspension at some elevation. The most erodible particles are in the size range of .004 inches diameter. Soil grains larger or smaller than this range are more resistant to movement. Smaller sizes tend to adhere to each other and larger particles have too much mass. Particles larger than a wheat grain usually are resistant to movement. When larger clods or rocks are present, they tend to disrupt the wind flow and protect the remaining soil.

Rough ground surface is the best deterrent to wind erosion. The turbulence caused by air friction reduces the velocity and tends to trap whatever particles are moving along the surface. The best defense is a standing crop. Where tall grass, or even crop stubble or plant litter is present, the effects of wind erosion are greatly reduced.

At locations where large expanses of loose soil are present, the use of tree rows between fields is at least somewhat effective in reducing soil movement by wind. Following the Dust Bowl days in the Great Plains of the central United States, many such tree rows were planted. Today, large quantities of soil may be seen accumulated along these rows.

Water erosion occurs whenever the rate of rainfall exceeds the surface storage capacity and infiltration rate of the soil. Soil is particularly vulnerable to erosion if it is not covered by a protective vegetative covering to absorb the impact of raindrops.

Surface storage is the result of rainfall accumulating in the small depressions that are common to all soil surfaces. These shallow depressions are capable of retaining a portion of most small rainstorms and holding the water until it either evaporates or infiltrates.

During intense storms (precipitation greater than 2 inches per hour) a significant portion of rainfall occurs as large drops which dislodge soil particles upon impact with bare soil. As the runoff travels downhill it carries the soil particles along.

From statistical observations, it has been shown that approximately 90 to 95% of soil erosion occurs during the most severe 2% of storms. Water erosion is the result of high intensity rainfall, steepness of slopes, length of unbroken slope (allows continued acceleration of water velocity), and lack of soil cover. Clear cutting of forested hillsides often results in severe erosion of soil.

During the early stages of water erosion, the upper layer of loosest soil (largely smaller grains) is washed off by the process of "sheet runoff." After the surface has been removed, small gullies are formed as the water follows the path of least resistance. Eventually, the small gullies coalesce into full-fledged severe gullies, and often have waterfalls and erode themselves to the headwaters of the watershed. The geological conclusion of a gully is reached after it has eroded itself downward to the base level which is typical of a stream or river.

Prevention of significant erosion can be accomplished by controlling the main causes. If a ground cover is maintained, the force of raindrops is reduced, generally to insignificant levels. When the slopes are reduced by such measures as contour ditching or if the length of steep slopes are minimized, water erosion is greatly reduced. The state, county, and federal agricultural agencies have prepared special erosion control programs for every area. Most of their assistance is very reasonably priced and well within the budget of all projects.

## CHAPTER 11

Management, Presentation and Interpretation of Soil Data

11.1 INTRODUCTION

Collection of site-specific information during an investigation is only part of the overall goal of an environmental project. Interpretation of the data as it relates to the project objectives is the heart of the project. Valid data interpretation is often a difficult task which requires that the investigator assume a combination role of Sherlock Holmes and Solomon the Wise. The best data possible have no value until they are applied.

This chapter focuses on several aspects of data management. Valid data presented in a suitable format often will result in a greatly reduced effort on the interpreter's task.

11.2 USING EXISTING SOURCES OF DATA

At the conceptual stages of a project, often as part of a proposal or preliminary plan, review of existing site background data presents the investigator with an opportunity to establish a general conceptual understanding of the site. Each bit of knowledge can improve the confidence factor of the project plan and reduce the number of unknown parameters. On the other hand, the preliminary data search may demonstrate that little local information is available and confirm the need for an extended field study.

In many cases, the project at hand is not entirely unique. Many sites have been studied in some detail as part of prior projects. Industrial or commercial sites probably have had foundation borings, sewer line excavations, water wells, or other earth penetrating excursions. If prior land use was primarily agricultural, the Soil Conservation Service has likely published a soil map along with its

detailed data. A preliminary listing of some potential sources of previously published soils data is presented below:

* U.S. Geological Survey reports (published and file data)
* Soil Conservation Service county soil surveys
* university libraries, including Masters and Doctorate studies
* seminar publications
* reports on file with state regulatory agencies
* documents on file with county or city building permits
* local water well drillers, engineers, and farmers
* computerized databases
* local high school and college instructors

11.3 RECOGNIZING THE VALUE OF DATA

Existing data collected by a diverse group of individuals for different purposes almost always have variable value. Soil descriptions found on a water
well drilling log form are valuable in that they define general formations (i.e.,
sand, clay, or rock), but not in sufficient detail for most environmental purposes. Conversely, when a general project is being performed, exceptionally detailed information can be confusing. An example of overly complex information is the study by a chemist with detailed mineralogical and physical analyses
made on every foot of soil core. The environmental user of the data may only
wish to determine the approximate vertical permeability for a hydrocarbon remediation. Clearly, the level of detail should be carefully considered. Several
general considerations should be employed when reviewing soils data from outside sources:

1. Who collected and evaluated the original data? Trained scientists and engineers
   generally provide more accurate information (but not always) than other sources.
   Analysis by a recognized laboratory may be more reliable than "self testing."
2. Which quality control standards were followed (if any)? Reports and studies which
   specify sample location, analytical procedures, duplicate/replicate/spike results probably hold more validity than others. References, photographs and clear maps are
   good indications.
3. What was the original purpose of the study? "Quick and dirty" reports may be reliable for very general information, but little more.
4. Were the data generated for litigation or prepared with a specific agenda? Data
   prepared for court use are generally very good; however, some advocate data are
   highly slanted toward a specific conclusion.
5. How and when were the data collected? Techniques of soil information collection
   have evolved over time. Analysis and sampling procedures used 25 years ago may
   have been described or reported according to different standards. Sample size,
   quantification limits, and testing procedures may have been different. An example is the abundance of chemical data prepared by "wet chemical" procedures.

While often quite valid, the lower detection limits were often .01% by weight, which is only to the nearest 100 ppm. Current detection limits are almost always <.1 ppm; more than 1000 times more sensitive.

6. Are the data internally consistent? Some comparative scrutiny can often be helpful. For example: the probability of plastic clay being in a state of compaction equal to 95% is very low. Fine grained sand seldom has a cation exchange capacity (CEC) which is greater than 10 meq/100 g.

7. Were consistent terms used by all the data sources? Various professionals over time have described the same phenomena in different terms. Particular attention should be given to local usage. Some terms, such as "hard pan," have distinctly different meanings in different locations. "Sandy clay" may or may not be equivalent to "clay loam" or SC.

8. Are the data original or "re-reported"? Many site-specific reports, especially initial reports, contain data which have been compiled from other reports; essentially rewritten data. After several renditions, errors or invalid interpretations have been known to become almost believable.

The assembly of soil data resulting from the preliminary studies, review of historical information, and recent field data collection often produces a wide variety of knowledge. Only after this compendium of data is accurately interpreted will characterization of the site be completed. Proper data management procedures are important to assure that each piece of available data is presented in a manner which allows interpretation in light of the other known facts.

Site data should be arranged and presented in a clear and logical format. Tabular, graphical, and other visual displays (i.e., contaminant isoconcentration maps) are essential for organizing and evaluating such data. Tables and graphs are not only useful for expressing results, but are also necessary for decision making during the field work. For example, a display of analytical results for each sampling location superimposed on a map of the site is helpful in identifying data gaps and in selecting future sampling locations. Graphs of concentrations of individual constituents plotted against the distance from a possible contamination source can help to identify patterns, which can be used to define future monitoring efforts.

Various tabular and graphical methods are available for data presentation. Table 11.1 presents a general summary of some of the most common data organization methods. The particular methods which are most applicable for a specific study are very project-specific, depending on the type of data, the topic being considered and other factors (such as the quantity of data available).

Often certain types of data, such as stratigraphy and sampling location coordinates, are more effectively displayed in graphical form. Such data may be presented in tabular form but should also be transformed into graphical presentations. For example, a cross section may be effectively illustrated on a two-dimensional cross sectional map. In some situations, such as in contamination migration studies, a three-dimensional characterization may be helpful.

Sample or date locations may be effectively illustrated on a topographic map. Topographic maps are an important ingredient for accurate data interpretation.

Table 11.1   Recommended Data Presentation Methods

Tables

Unsorted raw data
Sorted data

Graphical Formats and Other Visual Displays

Bar graphs
Line graphs
Area or plan maps
Isopleth (contour) plots
Groundwater flow nets
Cross-sectional plots, transects. or fence diagrams
Three-dimensional graphs

11.5 TABLES

Tabular presentations of both raw and sorted data are useful means of data presentation. Simple lists of data alone are not adequate to illustrate trends or patterns of data results. The presentation of these lists, however, serves as a good starting point for other presentation formats. These lists are also valuable for validation and auditing purposes. Tabular data should include the following information:

* Unique identification code
* Sampling location and sample type
* Source of data (who, what, why collected)
* Date of original generation
* Laboratory analysis number (if known)
* Property or component measured
* Result of analysis
* Detection limits
* Reporting units

The researcher should record all data, including outliers (data outside of the expected range) or data which are suspect. Potentially rejected data should be identified as such in the data tables, and explanation of why it should be rejected presented in the footnotes.

In addition to the resultant data, sampling logs should be maintained during the field investigation. Sampling logs are records of procedures used in taking environmental samples, and conditions prevailing at the site during sampling. Information contained in the log should include:

* Sampler identity
* Purpose of sampling
* Date and time of sampling

* Sample type and suspected contents (contaminated or not)
* Identification of sample location
* Sampling method, container type, and preservation
* Weight and volume of sample
* Number of samples collected
* Unique identification of sample
* Field observations
* Field measurements
* Weather conditions
* Identification of sampler

Probably the simplest form of data presentation is the organization of data into a format which displays trends or patterns. Examples of this type of presentation include medium tested, sampling date, sampling location, and constituent or property measured. Table 11.2 shows an example of a sorted table; data are sorted by medium, sampling date, and constituent measured. In Table 11.3 the data are sorted by medium, location, depth, and constituent analyzed. Inclusion of the sample identification number allows the reader to cross reference the data and look up any information not listed on the table.

Preparation of data summary tables can be simplified by the use of a computer spreadsheet program. These programs can perform sorting operations,

Table 11.2   Soil from Treatment Area B

| | | Concentration (ppm) | | | | |
Date	Location	Benzene	Toluene	Ethyl-benzene	Xylene (Total)	TPH[a]
6–7–92	5N-3W	120	750	360	983	8850
6–15–92	5N-3W	430	530	234	640	5700
6–27–92	8N-4W	11.0	410	152	415	2850
7–5–92	5N-6W	1.1	260	97.2	270	850

[a]Total Petroleum Hydrocarbon.

Table 11.3   Soil Analysis: Samples Collected Sept. 13, 1993

| | | | Concentration (ppm) | | |
Sample ID	Location	Depth	Pb	2,4 DNT	Hg
93–024–01	Boring No. 1	Surface	200	0.01	<0.2
93–024–02	Boring No. 2	6 inches	1900	0.01	<0.2
93–024–03	Boring No. 1	18 inches	50	0.05	<0.2
93–024–04	Boring No. 2	Surface	125	0.03	0.3
93–024–05	Boring No. 2	6 inches	700	0.07	0.5
93–024–06	Boring No. 2	18 inches	60	<0.01	<0.2
93–024–07	Boring No. 3	Surface	30	0.43	<0.2
93–024–08	Boring No. 4	18 inches	10	<0.01	0.5

perform simple calculations with the data, and display results in a number of tabular and graphical formats.

Graphical methods of data presentation will often illustrate trends and patterns better than tables. Some graphical formats that are useful for environmental data include bar graphs, line graphs, areal maps, and contour-plots.

Bar graphs and line graphs may be used to display changes in analyte concentrations by time, distance from a source, or other variables. Figure 11.1 com-

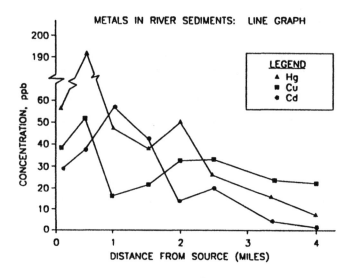

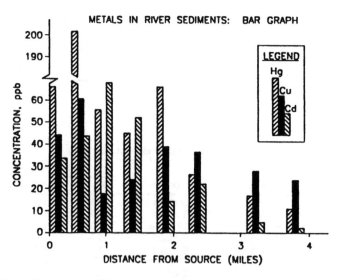

Figure 11.1   Comparison of line and bar graphs.

pares two methods of displaying changes of concentration over distance. Bar graphs are generally preferable to line graphs in instances where there is not enough information to assume continuity between data points. However, line graphs generally can display more information in a single graph.

The following principles of graphing should provide clear and effective line and bar graphs:

* Do not crowd data onto a graph. Plots with more than three or four lines or bar subdivisions become confusing. Different symbols or textures should be used to distinguish each line or bar.
* Choose the scale of the x and y axes so that data are spread over the full range of the graph. If one or two data points are far outside the range of the rest of the data, a broken line or bar may be used to indicate a discontinuous scale. If the data range exceeds two orders of magnitude, a logarithmic scale may be advantageous.
* The x and y axes of the plot should be clearly labeled with the parameter measured and the units of measurement.
* The x axis generally represents the independent variable and the y axis generally represents the dependent variable.

The distribution of constituents at a site may be represented by superimposing concentrations over a map of the site. Distributions may be shown by listing individual measurements, or by contour plots of the concentrations.

One approach is to plot the concentrations at discrete points. In this format, no assumptions are made concerning concentrations away from the immediate sampling area. For example, in Figure 11.2 soil PCB concentrations are shown by the height of the vertical bar at each sampling site. Soil samples indicated on this map were collected from approximately the same depth. Note that one bar is discontinuous so as to bring the lower values to a height that can be shown on the graph. Other representations of the same information could use symbols of different shapes, sizes, or colors to represent ranges of concentration. For example, a triangle might represent 0 to 10 ppb, etc.

Average concentrations surrounding sampling points may be shown by shading or texture patterns. Shading represents estimated areas of similar concentration only, and should not be interpreted as implying concentration gradients between adjacent points.

Lines of equal concentration are called isopleths (or isoconcentration). Construction of an isopleth map generally requires a relatively large number of sampling locations spaced regularly across the study area. An isopleth map is prepared by marking the site map with the concentrations detected at each sampling location. Lines connecting points of estimated equal concentrations are drawn according to the same principles used to construct elevation contours. Figure 11.3 demonstrates the use of an isopleth plot to demonstrate the distribution of contaminants at a site.

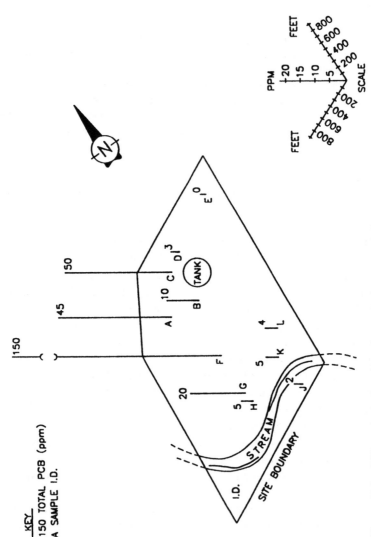

Figure 11.2   PCB concentrations in surface soils (ppm = mg/kg).

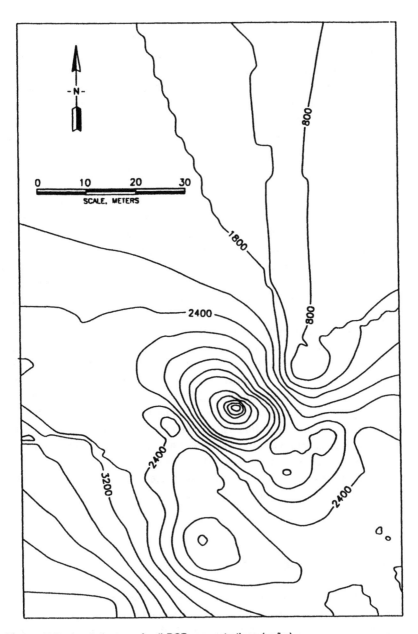

Figure 11.3 Isopleth map of soil PCB concentrations (µg/kg).

Isopach maps are a technique which is useful for displaying certain types of geological data. Isopachs are contour maps on which each line represents a unit of thickness of geological material (i.e., the thickness of a soil layer), as shown on Figure 11.4. This technique is useful if a material is known to be contained primarily within a highly permeable sand layer of varying thickness, confined between low permeability clay layers. The isopach map displays thickness, and does not provide information on absolute depth or slope.

Cross-sections or vertical profiles are particularly useful to display the distribution of an investigative parameter in all media. For soil and groundwater, the usual approach is to collect several soil cores which lie approximately in a straight line through the center of the area of concern. This cross section represents a transect of the site which displays subsurface stratigraphy, location of the boreholes, depth of investigation, location of the water table and other parameters as necessary. It is convenient to assemble the cross section so that the vertical data represent the actual elevation relationships between the borings. Concentrations of the material of concern may be indicated on the plot as discrete measurements. Two or more cross sections are often constructed across different parts of a site to demonstrate the distribution of vertical and horizontal analytes. Figure 11.5 presents a cross section used to demonstrate the vertical distribution of a petroleum contaminant.

A variation of cross sections is the fence diagram, which is often used when the boreholes are not in a straight line, or the geological setting is particularly complex. Because of the complexity of the data required to construct an accurate fence diagram, few are constructed for small environmental projects.

Computer graphics packages are available from several commercial suppliers which produce three-dimensional data plots. A common use of this technique is to represent concentrations across a study area as a three-dimensional surface, as shown in Figure 11.6. This type of graphic makes patterns in the data easier to visualize; however, precise concentrations cannot be displayed in this format because the apparent heights of contours change as the figure is rotated.

11.6 REPORTING OF OUTLIERS

Any real environmental measurement project will produce data which lie outside the "expected" range of values. Since field variability of environmental measurements can be great, determining whether an extreme (outlier) value is representative of actual levels may be difficult. Outliers may result from:

* A catastrophic unnatural (but real) occurrence, such as a spill
* Inconsistent sampling or analytical chemistry methodology
* Errors in the transcription of data values or decimal points, or
* True but extreme concentration measurements

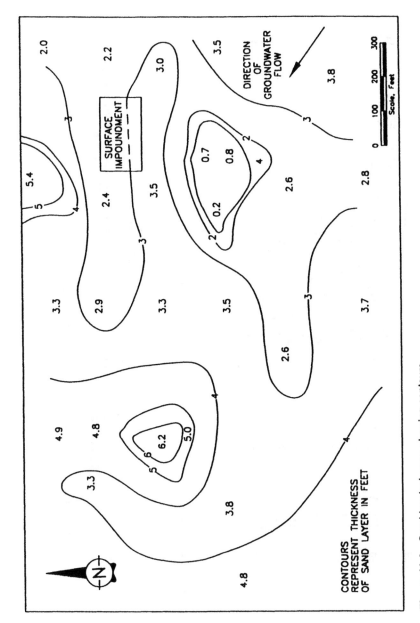

Figure 11.4 Sand isoplach map showing contours.

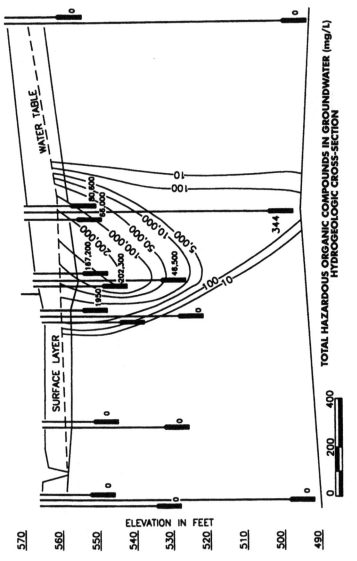

Figure 11.5 Transect showing concentration isopleths (µg/L).

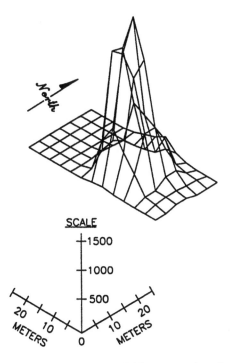

Figure 11.6 Three-dimensional data plot of soil PCB concentrations (µg/kg).

The data should be corrected if the outliers are caused by incorrect transcription, and the correct values can be obtained and documented. If a catastrophic event or a problem of methodology occurred (and can be confirmed), the data values should be reported with clear explanation reference. Documentation and validation of the cause of outliers must accompany any attempt to correct or delete data values because true, but extreme, values must not be altered. Statistical methods for identifying outliers require that the analytical laboratory have an ongoing program of quality assurance and that sufficient replicate samples be analyzed to account for field variability.

Outlier values should never be omitted from the raw data reported in the project tables. It is entirely possible that the outlier represents a heretofore unsuspected occurrence. Few environmental soils projects contain sufficient funding to assure a confidence level of 99+% accuracy.

11.8 REPORTING OF VALUES BELOW THE DETECTION LIMITS

Analytical values determined to be at or below the detection limit should be reported numerically (i.e., <.01 ppm), not assumed to be zero. Just because the analytical procedure used did not have the capability to detect below a specified concentration does not ensure that the analyte is not present.

## CHAPTER 12

Case Histories and Applications

The previous chapters have focused on the scientific and engineering aspects of soil science. Much of the discussion describes rather specific topics with limited reference to applications. This chapter presents a cross section of applied procedures for soil remediation. These cases are not intended for use as a "how-to" procedures manual, but rather to introduce concepts and to stimulate thoughtful consideration of possible procedures.

12.1 ADJUSTMENT OF SOIL pH

Blueberry plants thrive in soil with a pH between 4.5 and 5.5. While preparing a new bed, the gardener desired to optimize production. The soil in the bed is a fine silty sand, with at least 5% organic matter. Initial tests indicated that the average soil pH was near 7.0. Aluminum sulfate crystals were added in increments to the moist soil. After each acid treatment, the soil was thoroughly tilled. The pH was measured after 24 hours to allow time for chemical reactions to stabilize.

About 0.6 pounds of aluminum sulfate was required per cubic foot of topsoil to achieve pH 4.5. Continued testing and occasional addition of an acidification agent will probably be required to maintain the soil acid at this level.

While the quantity of aluminum sulfate required to meet the pH goal appears to be very large, this example demonstrates the logarithmic nature of pH adjustment. Soil which contains any significant quantity of organic matter and exchangeable ions has very high buffering capacity. Laboratory and field pilot testing are recommended prior to making large expenditures on remediation projects.

12.2 FLUSHING OF BRINE FROM SOIL

Following discovery of a ruptured oil field brine pipeline, investigators wished to evaluate cleanup alternatives. One option for remediation was to repair the line and allow natural rain infiltration to flush the brine from the soil. A laboratory experiment was performed to test the general feasibility of natural flushing.

An initial flushing experiment was made in a laboratory permeameter filled with clean medium sand (hydraulic conductivity 2.2×10^{-2} cm/sec). The sand-filled cylinder was saturated with potable water and vibrated (to increase the compacted density), then allowed to drain to simulate field capacity. Figure 12.1 describes the leaching column setup.

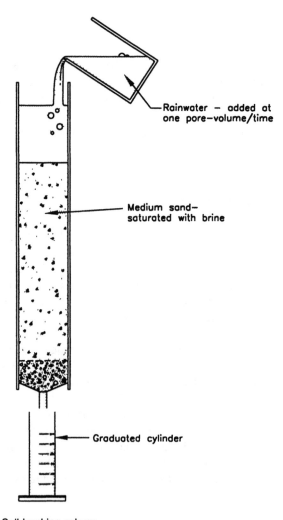

Rainwater — added at one pore-volume/time

Medium sand– saturated with brine

Graduated cylinder

Figure 12.1 Soil leaching column.

After several hours, a sample of the brine (40,500 ppm Total Dissolved Solids, TDS), equal to one pore space volume, was added to the cylinder and allowed to stand for approximately one hour.

Rainwater (<20 ppm TDS) was then poured through the column in increments of one pore-volume flush each. As the water seeped from the permeameter, periodic samples were collected for analysis.

Figure 12.2 is a graphical presentation of the flushing results. The initial discharge (collected during the addition of the brine) suggests that some soluble salts were present in the sand prior to the experiment. After the first pore volume exchange, the concentration of the discharge increased rapidly to a level approximately equal to one half that of the brine. After four flush volumes were exchanged, the concentration of salts in the discharged water approached that of the initial flush. The sixth volume of discharge water had TDS near that of the rainwater.

Results of this experiment indicate that brine flushing is a viable option for remediation. The major limitation to natural flushing is that the salts are only displaced by dilution. Downstream receivers of the diluted salt may not be receptive to degradation of their water quality.

In actual application, the soil will probably not be as permeable as the laboratory example, and may contain organic matter or other material which retains salt. It is not unreasonable that several more actual volume flushes may be required to reduce the salt concentration to acceptable levels.

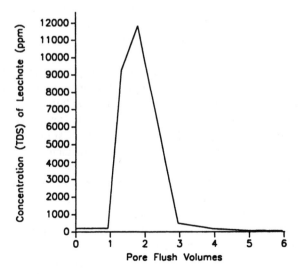

Figure 12.2 Brine flushing results.

12.3 SOIL VAPOR VENTING TEST FOR GASOLINE REMOVAL

A leaking underground fuel line at a service station was known to have released nearly 2000 gallons of gasoline into the subsurface. A site investigation
was made to define the aerial extent of contaminated soil and groundwater.

Geologically, the site is underlain by a loosely consolidated fine silty sand
to more than 100 feet. The water table is located at a depth of 10–13 feet with
less than 4 feet of seasonal fluctuation. Figure 12.3 is a cross section of the soil
venting site.

The initial remediation response was installation of several recovery wells
to retrieve free phase gasoline and contaminated groundwater. Vapor testing of
the unsaturated zone indicated that a significant quantity of gasoline was retained
on the soil. A pilot vapor extraction test was performed to determine if soil venting would be effective.

The vapor recovery test well was installed at a convenient location adjacent
to an existing monitoring well. The screened area of this well was set across the
vadose (unsaturated) zone. Four small diameter air pressure monitoring wells
(each 7.5 feet deep) were installed at distances of 2, 15, 25, and 45 feet from
the recovery well. Each air pressure monitoring well was equipped with a sensitive pressure differential gauge.

The vacuum pump used for this test was a ½ HP regenerative blower. A water
trap was installed between the blower and the vent well to separate condensed
water before it reached the blower unit. Figure 12.4 is a schematic diagram of
the test unit. Exhaust from the system was discharged to the atmosphere through
a tall vent pipe.

Throughout the 24 hour test, the performance was monitored continuously.
Air flow and exhaust temperature were measured by a commercial velocity/tem-
perature indicator installed in the exhaust vent pipe. The vacuum in the recovery well was monitored by a magni-helic gauge attached to the well head, and
the concentration of gasoline in the vented air was determined by a photoionization detector.

The vacuum on the vent well was continuously maintained at 55 inches of
water column (0.86 atmosphere), while the discharge rate was allowed to fluc-
tuate. Figure 12.5 presents a summary of test results. Initially, the discharge rate
was approximately 20 standard cubic feet per minute (scfm), with a gasoline recovery rate of 10×10^{-4} pounds per minute. As the test progressed, the rate of
gasoline recovery decreased and the rate of air flow increased.

Continuous measurements of the air pressure monitoring wells allowed determination of the radius of testing influence. After 24 hours, the 45 foot radius
well had a sustained 0.03 inches of water column vacuum, indicating that the
radius of influence extended at least that far.

During this test, only a small quantity of gasoline (approximately 1 pound)
was recovered; however, the test demonstrated that forced venting can result in
air flow through this geological unit. The initial gasoline recovery rate declined
quickly to an almost constant rate of 1×10^{-4} lb/min after the first 800 minutes.

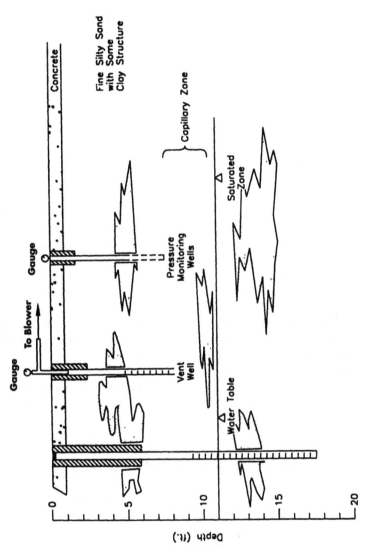

Figure 12.3 Subsurface conditions at soil venting test site.

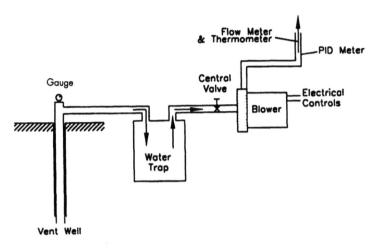

Figure 12.4   Schematic diagram of soil venting equipment.

This static rate indicates that the air was flowing through the soil too rapidly to establish concentration equilibrium. A slower pumping rate would result in higher recovery concentrations. Borings made subsequent to the test (within 15 feet of the recovery well) confirmed that the soil still contained relatively high residual gasoline concentration.

The steady increase in flow rate (from 20 to 38 scfm) resulted from an increase in air permeability due to dewatering of the unsaturated zone. The vacuum at the recovery well and the influx of relatively dry air are responsible for the continued increase of effective porosity. If the venting had continued, it is likely that the dewatering would have continued.

This test indicated that soil vapor venting at this site is an effective remediation technique. The combined processes of vapor removal and introduction of fresh air to stimulate bioactivity will be able to clean up this site at a relatively low cost.

12.4   SOIL WASHING TO REMOVE HEAVY METALS

(From "Soils Take a Bath at Superfund Site," by Jill Besch, in *Soils*, November 1993, pp. 22–25.)

A Superfund site in New Jersey was originally intended to be a recycling operation. During the early 1970s, chemical and metals processing companies brought their wastes there for processing. The recycling center failed to materialize and ownership reverted to the township in 1976 and the site was added to the Superfund list in 1985. Nearly 20,300 tons of contaminated soil were identified at the 10 acre site.

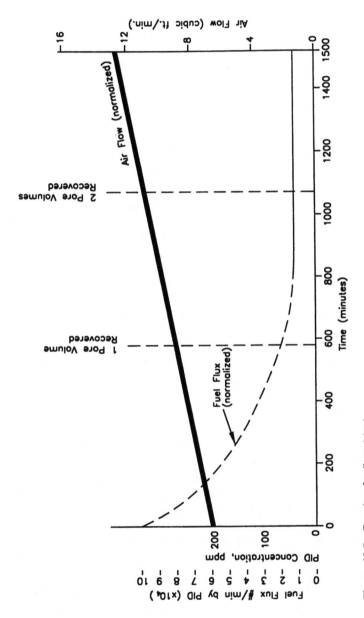

Figure 12.5 Results of soil vent test.

Several steps preceded the implementation of full-scale soil washing. The first step (early 1992) was to perform treatability and bench scale studies to define the particle size/contaminant relationship and design the treatment process train. Next, a demonstration run was made on 165 tons of actual soil. The pilot test successfully reduced contaminant concentrations to the target levels, which convinced regulatory officials to approve full-scale operation.

The preliminary testing revealed that most of the contaminants were associated with the finer particle sizes. Generally, site soils are classified into five primary fractions:

* Gross oversize: greater than eight inches and consisting of concrete rubble, tree stumps, branches, steel, tires
* Oversize: greater than two inches, less than eight inches, and consisting of cobbles, shredded wood, and slag
* Large coarse grain soils: in the range of 1/4 inch to two inches and consisting of sands and gravel
* Coarse grain soils: in the range of 40 to 60 microns up to 1/4 inch, consisting of sand
* Fine grain materials: silt and clay with an average particle size less than 40 to 60 microns

In June 1993, a larger scale pilot run cleaned 1000 tons of soils to levels well below the cleanup goals of 483 mg/kg for chromium, 3571 mg/kg for copper, and 1935 mg/kg for nickel. Full-scale remediation began in late June 1993.

The system is based on reducing the volume of materials to be disposed and returning as much clean soil as possible to the site. Typically, for every 100 tons of feed, an average of 85 tons of clean sand is recovered for reuse. Of the remaining 15%, about 10 tons is toxic waste and the remainder is rocks and pebbles.

The actual process begins with contaminated soils fed into a hopper and passed through a series of screens to eliminate the oversize material. Soils passing less than two inches are formed into a wet slurry and fed into a hydrocyclone, where the cut between fine grain and coarse grain material is made. The precise cut point for each specific site is determined during treatability studies. In the hydrocyclone, open to atmospheric pressure, centrifugal force causes the coarse grain sand to be spun out of the bottom, and the fine grain material and water to be ejected from the top of the unit. The hydrocyclones are the heart of the system, and can be field-adjusted precisely to meet the ultimate objective of the smallest possible sludge cake requiring off-site disposal.

The underflow from the hydrocyclone contains the coarse grain materials. Coarse materials are treated by a flotation technique whereby contaminants are concentrated in the froth, and the sands are dewatered and returned to the site as clean backfill. When treatment is required for this fraction, it is accomplished with air flotation treatment units. The air flotation tank is a long rectangular tank equipped with mechanical aerators and diffused air. Retention time is about 30 minutes. The choice of surfactants must be one that reduces the surface tension binding the contaminant to the sand and allows the contaminants to float into a froth, which is removed from the surface of the tank. The sands are dewatered and returned to the site as clean. The operator of the flotation unit must be experienced to control surfactant dose, slurry flow rate, air flow rate, and the height of the overflow weir.

Fines are forwarded to a sedimentation area consisting of banked lamella clarifiers (tall cylindrical tanks containing a series of baffles and chambers) where separation of the liquids and solids from the hydrocyclone overflow takes place.

Prior to introduction into the lamella clarifiers, soils are dosed with a polymer which has been selected during treatability testing. In the tank, the solids settle to the bottom and the liquids float on top. A slow moving mixer at the base of the tank prevents the solids from becoming lodged in the tank bottom. The thickened solids are then pumped out the bottom to a pressurized belt filter press. The solids influent is converted to a 45 to 55% dry solids filter cake. This cake contains the target contaminants and must be managed by regulated disposal at a permitted off-site facility, depending on the contaminants and their land ban status.

Product streams are tested and analyzed daily. The incoming feed pile is analyzed by computerized x-ray fluorescence (XRF) and an outside laboratory prior to introduction into the plant. During treatment, samples are retrieved from three streams. First, three samples of the process oversize are taken daily and analyzed on-site by a computerized XRF. A portion of each is preserved daily and combined into a weekly composite for analysis.

The oversize fraction usually tests clean. Clean sand is also sampled three times a day and analyzed on-site by XRF, composited weekly, and sent to the outside lab. Finally, the sludgecake, which contains the contaminants, is analyzed prior to off-site shipment.

The waste is also analyzed at the disposal facility according to the Toxicity Characteristic Leaching Procedure (TCLP) on the samples.

The washing system works best on soil composed of less than 40% clay or silt. It is *very* effective on sandy soils such as those found at this site. As the proportion of fine grain materials increases, the waste stream becomes more difficult to treat by soil washing only. In some cases, soil washing can be combined with other technologies, such as bioremediation, to complete the process train. Soil washing can remediate a wide range of contaminants, including heavy metals, polynuclear aromatic hydrocarbons, pesticides, and low level radioactive wastes. Since each site is unique, the arrangement of the equipment has to be selected based on the specific soil and contaminant characteristics.

12.5 SOIL REMEDIATION BY BIO-FIXATION

(Condensed from "Say It with Trees," by Edward Gatliff, in *Soils*, October 1993, pp. 16–18.)

Many people think in terms of microbes to bioremediate a contaminated site. Another approach enlists the aid of plants to assimilate contaminants. Known as phyto-remediation or agro-remediation, one company has developed the concept by utilizing trees as the biological agent. Under the trade name TreeMediation™, the process takes advantage of the extensive root system of trees and other plants to extract water from shallow aquifer systems.

Plant species can be selected to extract and assimilate or extract and chemically decompose target contaminants. Many inorganic chemicals considered environmental contaminants are, in fact, vital plant nutrients that can be absorbed through root systems for use in growth and development. Heavy metals can be taken up and bioaccumulated in plant tissues. Organic chemicals, notably pesticides, can be absorbed and metabolized by trees. The uptake of water can also substantially influence the local hydraulics of a shallow aquifer, thus controlling the migration of a contaminant plume. This pumping effect flushes water upward through the soil column and can be much more effective at remediation than traditional pump and treat systems. This technique has been used in aquifers up to 20 feet deep. Shallow, low yielding aquifers are particularly suited for this methodology.

Traditional pump and treat systems for this type of aquifer condition are often ineffective. The ideal mechanical pumping system is one in which a large number of wells are closely spaced and draw water upward through a soil column. This is what trees do.

Mechanical pumping appears to be effective in controlling the migration of contaminants in groundwater, but is unable to clean up the water to a pristine condition. One reason is that during pumping, channeling occurs where groundwater moves into the well, preferentially through higher permeability zones, thus having little effect on contaminants retained in less permeable soils. Another reason is soil bonding in which the contaminants chemically bond to constituents of the soil. TreeMediation is effective where these conditions exist.

As a passive method to control contaminant migration in shallow, low yield aquifers, this method is a low tech, environmentally compatible approach which requires very little maintenance.

Current projects focus on locations contaminated with chemicals from fertilizer handling and processing. Some of the sites also have pesticide and heavy metals problems.

Pesticide degradation is a well-documented process whereby plants and other organisms biochemically alter or degrade certain pesticides. At one site, surface soil with herbicide concentrations exceeding 1,000 ppm was reduced to less than 10 ppm using these principles.

A limitation to developing a system using plants to remediate heavy metals is that most vegetation won't appreciably accumulate the metals in its tissues. At sites where heavy metals are a concern, the use of trees acts as a hydraulic barrier (by water uptake) to leaching and off-site migration. Soluble metals are, however, often accumulated in the tissues of the trees.

A benefit of the long-term focus of this technology is that the low rate of uptake and accumulation of heavy metals is offset. It is irrelevant whether the process takes 100 days or 100 years when dealing with relatively immobile heavy metals. However, careful selection of plant species can result in effective and relatively rapid metals remediation.

At sites where no immediate threat to human health and the environment exists, bioremediation by plant fixation is an economical alternative to excavation and disposal.

12.6 BIODEGRADATION OF PETROLEUM CONTAMINATED SOILS

This bench scale study was conducted to evaluate the potential for biodegradation of petroleum hydrocarbons in soil samples from an abandoned refinery site. The objective was to determine if biodegradation by indigenous organisms could be used as a feasible remedial option.

A composite sample prepared from 22 separate soil samples was tested in this study. Water used for the experiment was groundwater retrieved from monitor wells at the site. In the laboratory, some preliminary analyses were performed (bacterial plate counts, pH, and existing nutrient concentration) to assess the modifications necessary to support bacterial growth during the testing.

The biodegradation study was performed in a series of electrolytic respirometers. These instruments provide a direct and continuous measurement of oxygen uptake by microbial populations within closed reaction vessels. Oxygen uptake by a microbial population is a direct measurement of biodegradation taking place within the reaction mixture.

Three sets of tests were conducted. Each set consisted of an identical quantity of soil mixed with a uniform quantity of site groundwater and nutrients. For the experiment, the following variations were used:

Set A: Nothing was added.
Set B: A small quantity of surfactant was added.
Set C: A small quantity of surfactant and a poison was added to allow determination of abiotic activity.

All reaction vessels were operated continuously under the same environmental conditions for the duration of the test. Samples removed from each after 0, 20, and 38 days were analyzed for petroleum hydrocarbons by capillary column FID-GC procedures.

Oxygen uptake in all nonpoisoned treatment vessels was immediate, indicating the presence of indigenous microorganisms previously adapted to the immediately available organic matter present. Oxygen uptake in the abiotic control was indicative of chemical oxidation of site matrix constituents.

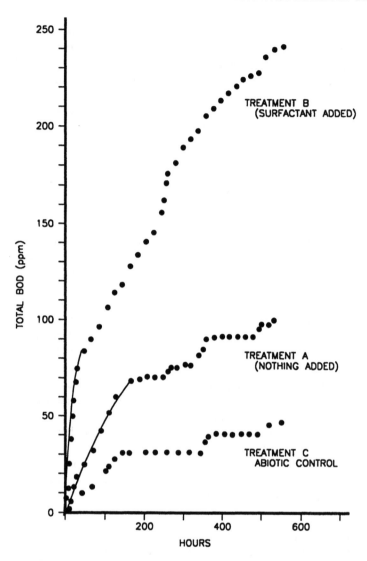

Figure 12.6   Natural microbial flora oxygen uptake for microbial growth in a liquid nutri-ent medium containing composite soil sample and ground water with (TREAT-MENT B) and without (TREATMENT A) surfactant.

Oxygen uptake in the treatment vessel without the surfactant was approximately 10 times lower than for the treatment vessel containing the surfactant. Ultimate biochemical demand (BOD) is defined as the total amount of oxygen required to biodegrade the immediately available organic matter present in a sample. Figure 12.6 presents the total BOD of each set of samples over time. Treatment B (with surfactant) was the most successful. Results of Set A indicated that the indigenous site bacteria were capable of biodegradation with addition of nutrients and oxygen. The GC analysis of the 20 and 38 day samples indicated that a wide variety of specific aliphatic and aromatic compounds were biodegraded during this test.

Results of this bench study were followed by a larger scale pilot test which was conducted to define practical design parameters. An actual field demonstration confirmed the feasibility of this technology.

12.7 SURFACE BIOREMEDIATION (LANDFARMING)

(From *Soils*, September-October 1991, p. 38.)

The basic principle of landfarming, also called surface bioremediation, is to enhance naturally-occurring degradation. Often, areas used for on-site surface bioremediation are considered to be temporary treatment areas which, after the treatment is completed, will be returned to original use.

Treatment time can range from 60 days to six months, depending on soil, volume contaminant, and cleanup goal. Lighter petroleum fractions (gasoline, diesel and heating oils) degrade faster than the heavier crudes.

Depending on site conditions and regulatory requirements, it may be necessary to install a liner beneath the bioremediation area. In the first step, contaminated soils are spread over the treatment area in a layer, usually six to 24 inches thick to allow as much oxygen as possible to reach the soil. The depth of the treatment zone should not exceed the capacity of disking equipment to expose the deepest sections to oxygen and nutrients.

After initial disking, phosphorous and nitrogen nutrients are added periodically. It is also possible to add bacteria enhancers to speed up the process. Then, periodic disking will increase exposure to oxygen and the other nutrients. Sites located in higher ambient temperatures degrade the hydrocarbons faster. Soils with low permeability recover more slowly than more permeable soils, and soils containing metals, salts, or chlorinated organics may be toxic to the added bacteria cultures. Some pesticides and herbicides may also compromise the performance of the bacteria.

One difficulty with the technique is that emissions control is difficult in areas where regulations restrict emissions. In the right situation, where permeable soil is contaminated with light hydrocarbons and where a suitable treatment site is conveniently located, landfarming has proved itself to be an effective and economical remediation technology.

12.8 RECYCLING CONTAMINATED SOIL INTO BRICK

(Condensed from "Turn Dirty Dirt into Brick," by Blaine Miller in *Soils*, August-September 1992, pp. 26–29.)

What soil recycling process can use petroleum contaminated soil to produce a contaminant-free product which can be sold for a profit? A company in North Carolina mixes petroleum contaminated soils with locally obtained clay and shale to manufacture brick.

The process blends clay and shale with the soils to form a plasticized mixture which is then extruded and molded into brick. Once the green (unfired) brick is dried and preheated, the kiln fires it at 1700 to 2000°F for approximately 12 hours. The temperature and residence time in the kiln destroy any organics and incorporate any inorganics into the vitrified brick product.

The contaminated soil is initially screened for debris and extensively blended with mined clay and shale into production stockpiles. The blending process is monitored and controlled to ensure production of feedstock which yields quality brick products. Grinders reduce this raw material to particles of acceptable size for brick formation. The raw material is mixed with water in a pug mill to increase plasticity.

The pug mill extrudes a continuous column of clay which is cut into green brick. These bricks are stacked on rail cars that can travel through tunnel kilns. Kiln travel time is approximately 2 ½ days. In the kiln, brick are first preheated to 600 to 1,600°F to remove moisture. At the peak temperature of 1,700 to 2,000°F, the kiln fires them for 12 hours. After cooling, the bricks are ready for packaging and shipment.

The brick manufacturing process can recycle various petroleum contaminated soil types including silts, sands, loams, and clays. This process is particularly suited for reuse of highly plastic clays that are difficult to treat with other remediation methods. Sandy soils also work well for their properties, which reduce firing shrinkage and improve moisture absorption from mortar—an important characteristic during brick laying.

In an interesting twist, some customers want their product back—in the form of brick. Several oil companies want to build their gas stations out of the recycled brick.

12.9 FIXATION OF PETROLEUM CONTAMINATED SOIL IN ASPHALT PAVING

(Condensed from "Paving Market Shows Promise," by Stephen M. Testa and Dennis L. Patton in *Soils*, November-December 1991, pp. 9–11.)

You pile petroleum contaminated soil on a road base. You mix it with liquid asphalt. You compact it with rollers. You've got a road. And you've eliminated the problems that tons of petroleum contaminated soil can cause. Or have you?

Several regulatory agencies are convinced that fixating contaminated soils in pavement is environmentally sound and cost effective, while providing minimal long-term liability. Some regulatory agencies are looking at ways to encourage this production for the nonhighway market.

The process works in this manner: the cold mix asphalt process uses an emulsifier to allow water to be combined with the asphalt to produce a mixture that has a viscosity low enough to mix with aggregate. This emulsion keeps asphalt particles separated from each other by a thin film of water. When the emulsified asphalt and aggregate have been mixed and placed, pressure from compaction breaks the water film to allow the asphalt particles to come into contact with each other and the aggregate.

The strength and durability of the pavement depends upon a mixture design that controls the type, size, and amount of aggregate. Generally, less than 10% of the total volume of an aggregate can be fines. This presents a major limitation for use of contaminated soil. If 5% of contaminated soil within the fines category is used in asphalt paving materials, 20 tons of final product must be manufactured to use up 1 ton of contaminated soil.

While highway-type paving materials are a substantial part of all asphalt paving materials produced, there are many additional uses that are not so closely controlled by rigorous specifications. Consequently, rather than attempt to meet rigorous specifications for highway-type pavement, the environment may be better served by producing a product of less strength and durability. Low-use road base, parking lots, dust abatement, bank stabilization, storage areas, and other such uses are all good candidates for paving material made with contaminated soil.

Field Methods for Soil Classification According to the USCS System

INTRODUCTION

This appendix provides practical field procedures for soil classification. Because most soil descriptions are made in the field, it is important that standard descriptive techniques be used. These procedures are based on those presented in the *Earth Manual*, published by the U.S. Bureau of Reclamation.

APPARATUS

Special apparatus or equipment is not required. However, the following items will facilitate the work:

1. A rubber syringe or a small oil can having a capacity of approximately 1/2 pint
2. A supply of clean water
3. A small bottle of dilute hydrochloric acid
4. A Classification Chart, Figure A-1.

PROCEDURE

The classification of a soil by this method is based on visual observations and estimates of its behavior in a remolded state. The procedure is, in effect, a process of elimination, beginning on the left side of the classification chart, Figure A.1 (see column headed Field Identification Procedures), and working to the right until the proper group symbol is obtained. The group symbol must be supplemented by a detailed word description, including a description of the in-place conditions.

UNIFIED SOIL CLASSIFICATION SYSTEM
Including Identification and Description

COARSE-GRAINED SOILS — More than half of material is larger than No. 200 sieve size

Note: The No. 200 sieve size is about the smallest particle visible to the naked eye

Classification	Subgroup	FIELD IDENTIFICATION PROCEDURES (Excluding particles larger than 3 inches and basing fractions on estimated weights)	GROUP SYMBOLS	TYPICAL NAMES
GRAVELS — More than half of coarse fraction is larger than No. 4 sieve size	CLEAN GRAVELS (Little or no fines)	Wide range in grain size and substantial amounts of all particle sizes	GW	Well-graded gravels, gravel-sand mixtures, little or no fines
		Predominately one size or a range of sizes with some intermediate sizes missing	GP	Poorly graded gravels, gravel-sand mixtures, little or no fines
	GRAVEL WITH FINES (Appreciable amount of fines)	Non-plastic fines (for identification procedures see ML below)	GM	Silty gravels, poorly graded gravel-sand-silt mixtures
		Plastic fines (for identification procedures see CL below)	GC	Clayey gravels, poorly graded gravel-sand-clay mixtures
SANDS — More than half of coarse fraction is smaller than No. 4 sieve size	CLEAN SANDS (Little or no fines)	Wide range of aggregates and substantial amounts of all intermediate particle sizes	SW	Well-graded sands, gravelly sands, little or no fines
		Predominately one size or a range of sizes with some intermediate sizes missing	SP	Poorly graded sands, gravelly sands, little or no fines
	SANDS WITH FINES (Appreciable amount of fines)	Non-plastic fines (for identification procedures see ML below)	SM	Silty sands, poorly graded sand-silt mixtures
		Plastic fines (for identification procedures see CL below)	SC	Clayey sands, poorly graded sand-clay mixtures

(For visual classification, the 1/4" size may be used as equivalent to the No. 4 sieve size.)

Notes: 1. Boundary classifications; soils possessing characteristics of two groups are designated by combinations of groups symbols, i.e., GW-GC well-graded gravel-sand mixture with clay binder.
2. All sieve sizes are U.S. standard.

Source: U.S. Bureau of Reclamation Earth Manual.

Figure A-1. Classification chart for coarse-grained soils.

UNIFIED SOIL CLASSIFICATION SYSTEM
Including Identification and Description

FIELD IDENTIFICATION PROCEDURES
(Excluding particles larger than 3 inches and basing fractions on estimated weights)

Identification Procedures on fraction smaller than No. 40 sieve size

Major Division		Dry Strength (Crushing Characteristics)	Dilatancy (Reaction to Shaking)	Toughness (Consistency Near Plastic Limit)	GROUP SYMBOLS	TYPICAL NAMES
FINE-GRAINED SOILS — More than half of material is smaller than No. 200 sieve size	SILTS AND CLAYS — Liquid limit less than 50	None to slight	Quick to slow	None	ML	Inorganic silts and very fine sands, rock flour, silty or clayey fine sands with slight plasticity
		Medium to high	None to very slow	Medium	CL	Inorganic clays of low to medium plasticity, gravelly clays, sandy clays, silty clays, lean clays
		Slight to medium	Slow	Slight	OL	Organic silts and organic silt-clays of low plasticity
	SILTS AND CLAYS — Liquid limit greater than 50	Slight to medium	Slow to none	Slight to medium	MH	Inorganic silts, micaceous or diatomaceous fine sandy or silty soils, elastic silts
		High to very high	None	High	CH	Inorganic clays of high plasticity, fat clays
		Medium to high	None to very slow	Slight to medium	OH	Organic clays of medium to high plasticity
HIGHLY ORGANIC SOILS		Readily identified by color, odor, spongy feel and frequently by fibrous texture			PT	Peat and other highly organic soils

Note: The No. 200 sieve size is about the smallest particle visible to the naked eye.

Figure A.2 Classification chart for fine-grained soils.

By briefly recording field observations made in the step-by-step procedure given below, the proper Unified Soil Classification of soil groups can be obtained.

Final field classification of soils for environmental projects is usually presented in the form of a log similar to Figure A.2. This particular log presents a graphic description, a detailed description, and a correlation of soil versus volatile organic content.

Note: Many natural soils will have properties not clearly associated with any one soil group, but which are common to two or more groups. Or, they may be near the borderline between two groups, exhibiting transition percentages of various grain sizes or plasticity characteristics. For this substantial number of soils, borderline classifications are used; an approximate dual symbol is assigned. A duel symbol consists of the two group symbols most nearly indicating the proper soil description connected by a hyphen, as for example, GW-GC, SC-CL, ML-CL, and others.

Selection and Preparation of a Sample

Select a representative sample of the soil and spread it on a flat surface or in the palm of the hand.

(a) Estimate and record the maximum particle size in the sample.

(b) Remove all particles larger than 3 inches from the sample. Estimate the percentage and distribution by weight (volume is satisfactory), of cobbles (3 to 12 inches in diameter) and boulders (particles over 12 inches in diameter). Remove the material, but record as descriptive information. *Only that fraction of the sample smaller than 3 inches is classified.*

Division Between Coarse- and Fine-Grained Soils

Classify the sample as coarse-grained or fine-grained by estimating the percent of particles which can be individually seen by the unaided eye. Soils containing more than 50% individually visible particles are coarse-grained soils. Soils containing less than 50% individually visible particles are fine-grained soils (see Figure A.1).

Note: For classification purposes, the No. 200 sieve size (0.074 mm) is the particle-size division between fine-grained and coarse-grained particles. Individual particles of this size are about the smallest that can be discerned by the unaided eye.

Visual Procedure for Coarse-Grained Soils

If it has been determined that the soil is coarse-grained, the soil is further identified by estimating and recording the percentage of: (1) gravel-sized particles, size range 3-inches to the No. 4 sieve (about 1/4 inch); (2) sand-sized particles, size range No. 4 sieve to No. 200 sieve; and (3) silt-sized and clay-sized particles, smaller than No. 200 sieve.

Note: The fraction of a soil smaller than the No. 200 sieve (clay and silt size) is referred to as "fines."

Gravelly Soils

If the percentage of gravel is greater than the sand, the soil is a GRAVEL, designated by the capital letter G. Gravel-sized particles are further divided as follows:

Coarse gravel - 3 inches to ¾-inch
Fine gravel - ¾-inch to No. 4 sieve (about ¼-inch)

Coarse or fine divisions are used to describe the average size of the gravel if poorly graded. Gravels are further identified as being CLEAN (when containing less than 5% fines), or DIRTY (when containing more than 12 % fines). However, the term "dirty" is usually not used in a description; instead, the properties of the fines that made the gravel "dirty" have to be described. Gravel containing 5 to 12% fines are given borderline classifications; that is, dual symbols. If the soil is obviously clean, the classification will be either:

1. GW, well-graded, if there is good representation of all particles sizes, or
2. GP, poorly-graded, if there is either predominant excess or absence of particle sizes within the gravel range. The letters W and P can be used in classification symbols for the coarse-grained soils only when the percentage of fines is less than 5%.

If the soil is obviously dirty, the classification will be either:

3. GM, if the fines have essentially no plasticity (silty), or
4. GC, if the fines are of low to medium or high plasticity (clayey).

Sandy-Soils

If the percentage of sand is greater than gravel, the soil is a SAND, designated by the capital letter S. Sand size particles are further divided as follows:

Coarse sand - No. 4 sieve (about 1/4-inch) to No. 10 sieve (about 3/32-inch)
Medium sand - No. 10 sieve (about 3/32-inch) to No. 40 sieve (about 1/64-inch)
Fine sand - No. 40 sieve (about 1/64-inch) to No. 200 sieve (about 3/1000-inch)

These divisions are used to describe the average size of the sand if poorly-graded.

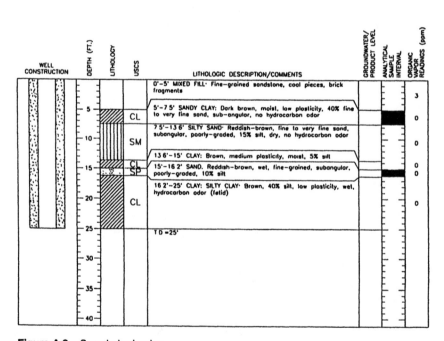

	TYPE	INTERVAL	MATERIAL	JOINT LENGTH	DIAMETER
CASING·	N/A				
SCREEN·	N/A				
GROUT·	Neet Cement	0' to 25'		AMOUNT:	
SEAL:	N/A				
FILTER PACK·	N/A				

SB-1

PAGE 1 OF 1

CLIENT· Example

PROJECT NO.: 94-032

DATE DRILLED: 5/3/94

SITE· Roll Cor Site

LOCATION Station N-93.0' & E-147.3'

DRILLING COMPANY· Clean Bore, Inc.

RIG· B-53

BOREHOLE: B-5o

LOGGED BY· L. G. Kennedy

DRILLING METHOD Hollow Stem Auger

FLUID: N/A

SAMPLING PROCEDURE· Visual

SAMPLING INTERVAL· Continuous

TOTAL DEPTH· 25'

DEVELOPMENT· N/A

NOTES· Boring at east margin of contamination (for confirmation)

ELEVATIONS: G L. 937.6   T.O C N/A   RISER HT. N/A

Figure A.3 Sample boring log.

The same procedure is applied as for gravels, except that the word SAND replaces GRAVEL, and the symbol "S" replaces "G." Thus, the clean sands will be classified as either:

1. SW or
2. SP,

and the dirty sands will be classified as:

3. SM, if the fines have minimal or no plasticity (silty), or
4. SC, if the lines are of low to medium or high plasticity (clayey).

Borderline Classifications for Coarse-Grained Soils

Borderline classifications can occur within the coarse-grained soil division, between soils within either the gravel grouping or the sand grouping, and between gravelly and sandy soils.

The procedure is to assume the coarser soil, when there is a choice, and complete the classification and assign the appropriate group symbol; then, beginning where the choice was made, assume the finer soil and complete the classification, assigning the second group symbol.

Borderline classifications within the gravel or sand groups can occur; symbols such as GW-GP, GM-GC, GW-GM, SW-SP, SM-SC, and SW-SM are common. Additionally, borderline classifications can also occur between the gravel and sand groups; soils classified as GW-SW, GP-SP, GM-SM, and GC-SC are common.

Borderline classifications can also occur within the fine-grained division, as well as between coarse- and fine-grained soils; classifications such as SM-ML and SC-CL are common.

Descriptive Information for Coarse-Grained Soils

The following information is required for a complete description of coarse-grained soils, and should be recorded. All of these descriptive data are not always needed. Judgment should be used to include pertinent information, to avoid negative information, and to eliminate repetition. However items 1, 2, 3, 8, and 11 should always be included. See Table A.1

1. typical name
2. maximum size distribution and approximate percentage of cobbles and boulders (particles larger than 3-inches) in the total material
3. approximate percentage of gravel, sand, and fines in the fraction of soil smaller than 3-inches
4. for poorly-graded materials, a statement of whether sand or gravel is coarse, medium, fine, or skip-graded

Table A.1   Checklist for Description of Coarse-Grained Soils

Items of Descriptive Data	Typical Information Desired for Sand and Gravel
Typical name _____	Gravel: Sand; Clayey GRAVEL; Silty SAND WITH COBBLES: (add descriptive adjectives for minor constituents—example: approximately 15% slight plasticity fines, medium toughness.)
Gradation _____	Well-graded: poorly-graded (uniformly-graded or skip-graded); (describe range of particle sizes, such as fine to medium sand or fine to coarse gravel, or the predominant size or sizes as coarse, medium, fine sand, or gravel.)
Size distribution _____	Approximate percent of gravel, sand, and fines in the fraction finer than 3 inches.
Plasticity of fines _____	None: low; medium; high.
Maximum particle size ___	Note the percent of boulders and cobbles (by volume), as well as the particle size.
Mineralogy _____	Rock hardness for gravel and sand. Note especially presence of mica flakes, shaly particles, organic matter, or friable particles.
Grain shape _____	Angular: subangular; subrounded; rounded.
Color _____	Use one basic color, if possible.
Odor _____	None; earthy; organic.
Moisture condition _____	Dry; moist; wet; saturated.
Degree of compactness ___	Loose; dense.
Structure _____	Stratified; lensed; nonstratified; heterogeneous.
Cementation _____	Weak; moderate; strong. Note reaction to HCl as: none; weak; moderate; or strong.
Local or geologic name ___	
Group symbol _____	GP, GW, SP, SW, GM, GC, SM, SC, or the appropriate dual symbol when applicable. Should be compatible with typical name used above.

  5. shape of the grains; rounded, subrounded, angular, subangular
  6. the surface coating, cementation, and hardness of the grains, and possible breakdown where compacted
  7. the color and organic content
  8. moisture conditions; dry, moist, wet, very wet (near saturation)
  9. plasticity of fines; none, slight, medium, high plasticity
  10. local or geologic name
  11. group symbol

Visual Procedure for Fine-Grained Soils

If it has been determined that the soil is fine-grained, the soil is further identified by estimating the percentages of gravel, sand, and fines (silt and clay-sized particles), and performing the manual identification tests for dry strength dilatancy and toughness (see field identification procedures for fine-grained soils and fractions in Figure A.2). By comparing the results of these tests with the requirements given for the six fine-grained soil groups, the appropriate group name and symbol is assigned. The same procedures are used to identify the fine-grained fractions of coarse-grained soils to determine whether they are silty or clayey.

Manual Identification Tests

The tests for identifying fine-grained soils (as opposed to "fines" included in soil) are performed on that fraction of the soil finer than the No. 40 sieve (about 1/64-inch).

The manual tests are considered to be performed on the "fines." The soil finer than the No. 40 sieve includes the "fines" (minus No. 200 sieve), and fine sand (minus No. 40 sieve to No. 200 sieve).

Select a small representative sample and remove by hand all particles larger than the No. 40 sieve size. Prepare two small specimens, each with a volume of about 1/2 cubic inch, by moistening until the specimens can easily be rolled into a ball. Perform the tests listed below, carefully noting the behavior of the soil pat during each test.

Note: Operators with considerable experience find that it is not necessary in all cases to prepare two pats. For example, if the soil contains dry lumps, the dry strength can be readily determined without preparing a pat for this particular purpose.

Dilatancy (Reaction to Shaking)

Add enough water to nearly saturate one of the soil pats. Place the pat in the open palm of one hand and shake horizontally, striking vigorously against the other hand several times. Squeeze the pat between the fingers. The appearance and disappearance of the water with shaking and squeezing is referred to as a reaction. This reaction is called: (1) quick, if water appears and disappears rapidly; (2) slow, if the water appears and disappears slowly; and (3) no reaction, if the water does not appear to change. Observe and record the type of reaction as descriptive information.

Toughness (Consistency Near Plastic Limit)

Dry the pat used in the dilatancy test (paragraph above), by working and molding until it has the consistency of putty. The time required to dry the pat is an indication of its plasticity. Roll the pat on a smooth surface or between the palms, into a thread about 1/8-inch in diameter. Fold and reroll the soil repeatedly into 1/8-inch diameter threads, so that its water content is gradually reduced until the 1/8-inch thread just crumbles. The water content at crumbling stage is called the plastic limit, and the resistance to molding at the plastic limit is called the toughness.

After the thread crumbles, the pieces should be lumped together and a slight kneading action continued until the lump crumbles. If the lump can still can be molded slightly dryer than the plastic limit and if high pressure is required to roll the thread between the palms of the hands, the soil is described as possessing high toughness. Slight toughness is indicated by a weak thread that breaks easily and cannot be lumped together when drier than the plastic limit. Medium toughness is assigned to those soils which fall somewhere between high and slight toughness characteristics. This test also provides approximate information on the plasticity index, PI, of the soil. The number of times the procedure can be repeated is an indication of the PI of the material.

Highly organic clays have a very weak and spongy feel at the plastic limit. Nonplastic soils cannot be rolled into a thread of 1/8-inch diameter at any water content. Observe and record the toughness as descriptive information.

Dry Strength (Crushing Resistance)

Completely dry one of the prepared specimens. Then, measure its resistance to crumbling and powdering between the fingers. This resistance, called dry strength, is a measure of the plasticity of the soil and is influenced largely by the soil's colloidal fraction. The dry strength is designated as slight if the dried pat can be easily powdered, medium if considerable finger pressure is required, and high if it cannot be powdered at all. Observe and record the dry strength as descriptive information.

Note: The presence of high-strength, water-soluble cementing materials, such as calcium carbonates or iron oxides, may cause high dry strengths. Nonplastic soils, such as cliche, coral, crushed limestone, or soils containing carbonaceous agents, may have high dry strengths, but this can be detected by the effervescence caused by the application of dilute hydrochloric acid (see Acid Test).

Organic Content and Color

Fresh, wet organic soils usually have a distinctive odor of decomposed organic matter. This odor can be made more noticeable by heating the wet sample. Another indication of the organic material is the distinctive dark color. Dry, inorganic clays develop an earthy odor upon moistening which is distinctive from that of decomposed organic matter.

Other Identification Tests

1. Acid Test: The acid test using dilute hydrochloric acid (HCl) is primarily a test for the presence of calcium carbonate. For soils with high dry strength, a strong reaction indicates that the strength may be due to calcium carbonate

as cementing agent, rather than colloidal clay. The results of this test (reaction to HCl), should be reported in the soil description.

Note: dilute solution (1:3) of hydrochloric acid is one part of concentrated hydrochloric acid to three parts of distilled water. Handle with caution. Rinse with tap water if it comes in contact with skin.

2. Shine: This is a quick supplementary procedure for determining the presence of clay. The test is performed by cutting a lump of dry or slightly moist soil with a knife. A shiny surface imparted to the soil indicates highly plastic clay, while a dull surface indicates silt or clay of slight plasticity.
3. Miscellaneous: other criteria undoubtedly can be developed as the individual investigator gains experience in classifying soils. For example, differentiation between some of the fine-grained soils depends largely upon the experience in the "feel" of the soils. Frequent checking by laboratory tests is necessary to gain this experience.

Silty and Clayey Soils

Various combinations of results from manual identification tests indicate which grouping is proper for the soil in question.

(a) The following three groups are soils possessing slight to medium plasticity (symbol L):

1. ML has little or no plasticity, and may be recognized by slight dry strength, quick dilatancy, and slight toughness.
2. CL has slight to medium plasticity, and may be recognized by medium to high dry strength, very slow dilatancy, and medium toughness.
3. OL is less plastic than the clay (CL), and may be recognized by slight to medium dry strength, medium to slow dilatancy, and slight toughness. Organic matter must be present in sufficient amount to influence the soil properties in order for a soil to be placed in this group.

(b) The following three groups are soils possessing slight plasticity to high plasticity (symbol H):

1. MH is generally very absorptive. It has slight to medium plasticity, and may be recognized by low dry strength, slow dilatancy, and slight to medium toughness. Some inorganic soils (such as kaolin, which is a clay from a mineralogical standpoint) possessing medium dry strength and toughness will fall in this group.
2. CH possesses high plasticity, and may be recognized by high dry strength, no dilatancy, and usually high toughness.
3. OH is less plastic than the fat clay (CH), and may be recognized by medium to high dry strength, slow dilatancy, and slight to medium toughness. Organic matter must be present in sufficient amount to influence soil properties in order for soil to be placed in this group.

Table A.2   Checklist for Description of Fine-Grained and Partly-Organic Soils

Items of Descriptive Data	Typical Information Desired for Sand and Gravel
Typical name _____	SILT: sandy SILT[a]; CLAY; lean or fat CLAY; sandy CLAY; silty CLAY; organic SILT; organic CLAY
Size distribution _____	Approximate percent of fines, sand, and gravel fraction less than 3 inches in size. Must add to 100 percent.
Plasticity of fines _____	None; low; medium; high.
Dry strength _____	None; low; medium; high.
Dilatancy _____	None; very slow; slow; medium; quick.
Toughness near _____ plastic limit	None; slight (low); medium; high.
Maximum particle size ____	Note percentage of cobbles and boulders (by volume), as well as maximum particle size.
Color _____	Use one basic color, if possible. Note percentage of mottling or banding.
Odor _____	None; earthy; organic.
Moisture condition _____	Dry; moist; wet; saturated.
Consistency _____	Very soft; soft; firm; hard; very hard.
Degree of compactness ___	Loose; dense.
Structure _____	Stratified; aminated (varved); fissured; blocky; lensed; homogeneous.
Cementation _____	Weak; moderate; strong. Note reaction to HCl as: none; weak; moderate; or strong.
Local or geologic name ___	
Group symbol _____	CL, CH, ML, OL, OH, Pt, or the appropriate dual symbol when applicable. Should be compatible with typical name used above.

[a]25% or more sand must be present. "Gravelly" can be substituted for "sandy" where applicable. Includes cobbles and boulders in typical name where applicable.

(c) Borderline classifications can occur within the fine-grained soil division, between low and high liquid limit soils, and between silty and clayey soils. The procedure is comparable to that given for coarse-grained soils above; that is, first assume a coarse soil, when there is a choice, then a finer soil, and assign dual group symbols.

Borderline classifications which are common are as follows: ML-MH, CL-CH, OL-OH, CL-ML, ML-OL, CL-OL, MH-CH, MH-OH, and CH-OH.

Peat or Very Highly Organic Soils (Symbol Pt)

These may be readily identified by color, odor, sponginess or fibrous texture.

Descriptive Information for Fine-Grained Soils

The following information is required for a complete description of fine-grained soils and should be recorded. All of these descriptive data are not always needed. Judgment should be used to include pertinent information, to avoid negative information, and to eliminate repetition. However, items 1, 2, 6, 7, and 9 should always be included. See Table A.2.

1. typical name
2. maximum particle size. Distribution, and approximate percentage of cobbles and boulders (particles larger than 3 inches) in the total material
3. approximate percentage of gravel, sand, and fines in the fraction of soil smaller than 3 inches
4. hardness of the coarse grains, possible breakdown into smaller sizes
5. color in moist condition and organic content
6. moisture and conditions; dry, moist, wet, very wet (near saturation)
7. plasticity characteristics; none, slight, medium, high plasticity
8. local or geologic name
9. group symbol

Glossary of Soil Terms

absorption: The process by which one substance is taken into and included within another substance, as the absorption of water by soil or nutrients by plants.

acid soil: Soil with a pH value < 7.0.

acidity, total: The total acidity in a soil or clay. Usually it is estimated by a buffered salt determination of (cation-exchange minus exchangeable bases) = total acidity.

actinomycetes: A nontaxonomic term applied to a group of gram-positive bacteria that have a superficial resemblance to fungi. Includes many, but not all, organisms belonging to the order Actinomycetales.

adsorption: The increased concentration of molecules or ions at a surface, including exchangeable cations and ions on soil particles.

adsorption isotherm: A graph of the quantity of a given chemical species bound to an adsorption complex, at a fixed temperature, as a function of the concentration of the species in a solution that is in equilibrium with the complex.

aeration, soil: The process by which air in the soil is replaced by air from the atmosphere.

aerobic: (1) Having molecular oxygen as a part of the environment; (2) growing only in the presence of molecular oxygen, as aerobic organisms; (3) occurring only in the presence of molecular oxygen.

aggregate: A unit of soil structure, usually formed by natural processes in contrast with artificial processes and generally <10 mm in diameter.

aggregation: the act of soil particles cohering so as to behave mechanically as a unit.

agronomy: The branch of agriculture which deals with the theory and practice of field-crop production and soil management.

air-dry: (1) The state of dryness (of a soil) at equilibrium with the moisture content in the surrounding atmosphere. The actual moisture content will depend upon the relative humidity and the temperature of the surrounding atmosphere; (2) to allow to reach equilibrium in moisture content with the surrounding atmosphere.

air porosity: The fraction of the bulk volume of soil that is filled with air at any given time or under a given condition, such as a specified soil-water content or soil-water matric potential.

Alfisols: An order of the U.S. system of soil taxonomy. Mineral soils that have umbric or ochric epipedons, argillic horizons, and that hold water at < 1.5 MPa tension during at least 90 days when the soil is warm enough for plants to grow outdoors. Alfisols have a mean annual soil temperature of < 8°C, or a base saturation in the lower part of the argillic horizon of 35% or more when measured at pH 8.2.

alkaline soil: Any soil having a pH of > 7.0.

alluvial: Pertaining to processes or materials associated with transportation or deposition by running water.

ammonia fixation: Chemisorption of ammonia (NH_3), and possibly ammonium, by the organic fraction of the soil.

ammonium phosphate: a generic class of phosphate fertilizers.

amorphous material: noncrystalline solid mineral constituents of soil.

anaerobic: (1) The absence of molecular oxygen; (2) growing in the absence of molecular oxygen (i.e., anaerobic bacteria); (3) occurring in the absence of molecular oxygen (as a biochemical process).

anaerobic respiration: The metabolic process whereby electrons are transferred from an organic compound to an inorganic acceptor molecule other than oxygen. The most common acceptors are carbonate, sulfate, and nitrate.

anion exchange capacity: The sum total of exchangeable anions that a soil can adsorb.

apparent cohesion: Cohesion in granular soils due to capillary forces associated with water.

argillic horizon: A mineral soil horizon that is characterized by the illuvial accumulation of layer-lattice silicate clays.

aridic: A soil moisture regime that has no water available for plants for more than half the cumulative time that the soil temperature at 50 cm below the surface is >5°C, and has no period as long as 90 consecutive days when there is water for plants while the soil temperature at 50 cm is continuously > 8°C.

Aridisols: An order in the U.S. system of soil taxonomy. Mineral soils which have an aridic moisture regime, an ochric, epipedon, and other pedogenic horizons but no oxic horizon.

Atterberg Limits: The collective designation of so-called limits of consistency of fine-grained soils suggested by Albert Atterberg. Usually presented as the liquid limit (LL), plastic limit (PL), and the plasticity index (PI).

available water: The portion of water in a soil that can be readily absorbed by plant roots. Considered by most workers to be that water held in the soil against a pressure of up to approximately 15 bars.

bar: A unit of pressure equal to one million dynes per square centimeter.

base-saturation percentage: The extent to which the adsorption complex of a soil is saturated with exchangeable cations other than hydrogen. It is expressed as a percentage of the total cation-capacity.

bearing capacity: Ability of a material to support a load normal (perpendicular) to the surface.

bedrock: The more or less continuous body of rock which underlies the overburden soils.

bentonite: An expansive clay formed from the decomposition of volcanic ash.

breakthrough curve: The relative solute concentration in the outflow from a column of soil or porous medium after a step change in solute concentration has been applied to the inlet end of the column, plotted against the volume of outflow (often in number of pore volumes).

buffer compounds, soil: The solid and solution phase components of soils that resist appreciable pH change in the soil solution, i.e., carbonates, phosphates, oxides, phyllosilicates, and some organic materials.

bulk density, soil: The mass of dry soil per unit bulk volume. The bulk volume is determined before drying to constant weight at 105°C.

bulk specific gravity: The ratio of the bulk density of a soil to the mass of unit volume of water.

bulk volume: The volume, including the solids and the pores, of an arbitrary soil mass.

calcareous soil: Soil containing sufficient $CaCO_3$ and/or $MgCO_3$, to effervesce visibly when treated with cold dilute HCl.

caliche: A zone near the surface, more or less cemented by secondary carbonates of Ca or Mg precipitated from the soil solution. It may occur as a soft thin soil horizon, as a hard thick bed or as a surface layer exposed by erosion.

capillary attraction: A liquid's movement over or retention by a solid surface due to the interaction of adhesive and cohesive forces.

capillary fringe: A zone just above the water table (zero gauge pressure) that remains almost saturated. The extent and degree of definition of the capillary fringe depends upon the size-distribution of the pores.

capillary migration (capillary flow): The movement of water (or other liquid) by capillary action.

carbon-organic nitrogen ratio: The ratio of the mass of organic carbon to the mass of organic nitrogen in soil, organic material, plants, or the cells of microorganisms.

cation-exchange: The interchange between a cation and solution and another cation on the surface of any surface-active material such as a clay colloid or organic colloid.

cation-exchange capacity (CEC): The sum total of exchangeable cations that a soil can absorb. Expressed in milli-equivalents per 100 grams or gram of soil (or other exchangers such as clay).

cemented: Indurated; having a hard, brittle consistency because the particles are held together by cementing substances such as humus, $CaCO_3$, or the oxides of silicon, iron, and aluminum. The hardness and brittleness persist even when wet.

chemical weathering: The breakdown of rocks and minerals due to chemical activity, primarily due to the presence of water and components of the atmosphere.

chroma: The relative purity, strength, or saturation of a color, directly related to the dominance of the determining wavelength of the light and inversely related to grayness; one of the three variables of color.

clay: (1) A soil particle consisting of particles > 0.002 mm in equivalent diameter; (2) a textural class.

clay mineral: Naturally occurring inorganic crystalline material found in soils and other earthy deposits, the particles being clay sized; that is, > 0.002 mm in diameter.

clod: A compact, coherent mass of soil ranging in size from 5 to 100 mm to as much as 20 to 25 cm, and produced artificially, usually by the activity of man and plowing, digging, etc., especially when these operations are performed on soils that are either too wet or too dry for normal tillage operations.

coarse fragments: Rock or mineral particles > 2.0 mm in diameter.

coarse texture: The texture exhibited by sand, loamy sands, and sandy loams, except very fine sandy loams.

COD (chemical oxygen demand): A measure of oxygen-consuming capacity of inorganic and organic matter present in water or waste water. It is expressed as the amount of oxygen consumed from a chemical oxidant in a specific test.

cohesion: The force holding a solid or liquid together, owing to attraction between like molecules.

cohesionless soils: A soil that, when unconfined, has little or no strength when air-dried, and that has little or no cohesion when submerged in water.

cohesive soil: A soil that, when unconfined, has considerable strength when air-dried and that has significant cohesion when submerged.

colloidal particles: Particles that are so small that the surface has an appreciable influence on the properties of the particle.

compaction: The densification of a soil by means of mechanical manipulation.

compaction curve (Proctor curve) (moisture-density curve): The curve showing the relationship between the dry unit weight (density) and the water content of a soil for a given compactive effort.

compaction test (moisture-density test): A laboratory compacting procedure whereby a soil at a known water content is placed in a specified manner into a mold of given dimensions, subjected to a compactive effort of controlled magnitude, and the resulting unit weight is determined. The procedure is repeated for various water contents sufficient to establish a relationship between water content and unit weight.

compressive strength (confined or uniaxial compressive strength): The load per unit area at which an unconfined cylindrical specimen of soil or rock will fail in a simple compression test. Commonly the failure load is the maximum that the specimen can withstand in the test.

cone penetrometer: An instrument in the form of a cylindrical rod with a cone-shaped tip designed for penetrating soil and for measuring the end-bearing component of penetration resistance. The resistance to penetration developed by the cone equals the vertical force applied to the cone, divided by its horizontal projected area.

consistency: The resistance of a material to deformation or rupture. Also, the degree of cohesion or adhesion of the soil mass. Engineering descriptions include: soft, firm or medium, stiff, very stiff or hard.

Darcy's law: (1) A law describing the rate of flow of water through porous media. (Named for Henry Darcy of Paris who formulated it in 1856 from extensive work on the flow of water through sand filter beds.) As formulated by Darcy, the law is:

$$Q = \frac{kS(H+e)}{e}$$

where:

Q = the volume of water passed in unit time,
S = the area of the bed,
e = the thickness of the bed,
H = the height of the eater on top of the bed, and
"k = a coefficient depending on the nature of the sand and the weight of the atmospheres"

(2) Generalization for three dimensions: The rate of viscous flow of water in isotropic porous media is proportional to, and in the direction of, the hydraulic gradient.

(3) Generalizations for other fluids: The rate of viscous flow of homogenous fluids through isotropic porous media is proportional to, and in the direction of, the driving force.

degradation: The breakdown of substances by biological action.

degree of saturation: The extent or degree to which the voids in soil or rock contain fluid (water, gas, or oil). Usually expressed in percent related to total void or pore space.

desorption: The displacement of ions from the solid phase of the soil into solution by a displacing ion.

diffusion: The movement of a solute in soil or groundwater that results from a concentration gradient.

effective stress: The stress transmitted through a soil by intergranular pressures.

Eh: The potential that is generated between an oxidation or reduction half-reaction and the H electrode in the standard state.

Entisols: An order in the U.S. system of soil taxonomy. Mineral soils that have no distinct subsurface diagnostic horizons within 1 m of the soil surface.

eolian: Pertaining to material transported and deposited by the wind. Includes earth materials ranging from dune sands to silty loess deposits.

essential chemical elements: Elements required by plants to complete their life cycles.

evaporites: Residue of salts (including gypsum and all more soluble species) precipitated by evaporation.

evapotranspiration: The combined loss of water from a given area, and during a specified period, by evaporation from the soil surface and by transpiration from plants.

exchange capacity: The total ionic charge of the adsorption complex active in the adsorption of ions.

exchangeable anion: A negatively charged ion held on or near the surface of a solid particle by a positive surface charge, and which may be replaced by other negatively charged ions.

exchangeable cation: A positively charged ion held on or near the surface of a solid particle by a negative surface charge of a colloid, and which may be replaced by other positively charged ions in the soil solution.

facultative organism: An organism that is capable of both aerobic and anaerobic metabolism.

fertilizer: Any organic or inorganic material of natural or synthetic origin that is added to a soil to supply one or more elements essential to the growth of plants.

fertilizer grades: The percent proportions of primary nutrients in a commercial fertilizer expressed as percent of N-P-K, in that order. Examples are 10-10-10, 5–10–5, and 20–10–10.

field capacity (field moisture capacity): The percentage of water remaining in a soil 2 or 3 days after having been saturated and after free drainage has practically ceased.

film water: A layer of water surrounding soil particles, and varying from 1 or 2 to perhaps 100 or more molecular layers. Usually considered as that water remaining after drainage has occurred because it is not distinguishable in saturated soils.

fine texture: Consisting of or containing large quantities of the fine fractions, particularly of silt and clay. (Includes all clay loams, sandy clay loam, sandy clay, silty clay, and clay textural classes. Sometimes subdivided into clayey texture and moderately fine texture.)

flux density: The time rate of transport of a quantity (i.e., mass or volume of fluid, electromagnetic energy, number of particles, or energy) across a unit area perpendicular to the direction of flow.

frost heave: The lifting or lateral movement of soil caused by freezing of water resulting in the formation of ice lenses or ice needles.

free water (gravitational water, groundwater): Water that is free to move through a soil or rock mass under the influence of gravity.

glacial drift: Rock debris which has been transported by glaciers and deposited, either directly from the ice or from the melt-water. The debris may or not be heterogeneous.

grading: A "well-graded" sediment containing some particles of all sizes in the range concerned. Distinguished from "well-sorted," which describes a sediment with grains of one size.

grain-size analysis (mechanical analysis or particle-size analysis): The process of determining grain-size distribution.

groundwater: The portion of the total precipitation which at any particular time is either passing through or standing in the soil and the underlying strata, and is free to move under the influence of gravity.

gypsum: The common name for calcium sulfate ($CaSo_4 \cdot H_2O$), used to supply calcium and sulfur and to ameliorate sodic soils.

hardpan: A hardened soil layer, in the lower A or in the B horizon, caused by cementation of soil particles with organic matter or with materials such as silica, sesquioxides, or calcium carbonate. The hardness does not change appreciably with changes in moisture content, and pieces of the hard layer do not slake in water.

heterogeneity: Having different properties at different points.

Histosols: An order in the U.S. system of soil taxonomy. Organic soils that have organic soil materials in more than half of the upper 80 cm, or that are of any thickness, if overlying rock or fragmental materials that have interstices fill with organic soil materials.

hue: One of the three variables of color. It is caused by light of certain wavelengths, and changes with the wavelength.

humic acid: The dark-colored organic material that can be extracted from soil by various reagents (i.e., dilute alkali) and is precipitated by acidification to pH 1 or 2.

humus: (1) Total of the organic compounds in soil exclusive of undecayed plant and animal tissues, their "partial decomposition" products, and the soil biomass. The term is often used synonymously with "soil organic matter;" (2) organic layers of the forest floor.

hydration: The physical binding of water molecules to ions, molecules, particles, or other matter.

hydraulic gradient: The loss of head pressure per unit distance of flow.

hydric soils: Soils that are wet long enough to periodically produce anaerobic conditions, thereby influencing the growth of plants.

hydrodynamic dispersion: The process wherein the solute concentration in flowing solutions changes in response to the interaction of solution movement with pore geometry of the soil, a behavior with similarity to diffusion but only taking place when solution movement occurs.

hydrogen bond: The chemical bond between a hydrogen atom of one molecule and two unshared electrons of another molecule.

hysteresis: A unique relationship between two variables, wherein the curves depend on the sequences or starting point used to observe the variables.

Examples include the relationships: (1) between soil-water content and soil-water matric potential; and (2) between solution concentration and adsorbed quantity of chemical species.

hydrostatic pressure: A state of stress in which all the principal stresses are equal (and there is no shear stress).

hygroscopic water: Water adsorbed by a dry soil from an atmosphere of high relative humidity, water remaining in the soil after "air-drying," or water held by the soil when it is in equilibrium with an atmosphere of a specified relative humidity at a specified temperature, usually 98% of relative humidity at 25°C.

igneous rock: Rock formed from the cooling and solidification of magma, and that has not been changed appreciably since its formation.

immobilization: The conversion of an element from the inorganic to the organic form in microbial tissues or in plant tissues.

Inceptisols: An order in the U.S. system of soil taxonomy. Mineral soils that have one or more pedogenic horizons in which mineral materials other than carbonates or amorphous silica have been altered or removed but not accumulated to a significant degree. Under certain conditions, Inceptisols may have an ochric, umbric, histic, plaggen, or mollic epipedon. Water is available to plants more than half of the year or more than 90 consecutive days during a warm season.

infiltration: The downward entry of water into the soil.

intrinsic permeability: The property of a porous material that expresses the ease with which gases or liquids flow through it. Often symbolized by:

$$k = \frac{Kn}{pg}$$

where:

K = the Darcy hydraulic conductivity
n = viscosity of the fluid
p = fluid density and,
g = acceleration of gravity.

ions: Atoms, groups of atoms, or compounds which are electrically charged as a result of the loss of electrons (cations) or the gain of electrons (anions).

ion exchange: A chemical process involving reversible interchange of ions between a liquid and a solid but no radical change in structure of the solid.

isotropic: Having the same properties in all directions.

kaolin: A subgroup name of aluminum silicates with a 1:1 layer structure. Kaolinite is the most common clay mineral in the subgroup.

laminar flow: Flow in which the head loss is proportional to the first power of the velocity.

leaching: The removal of materials in solution from the soil.

liquefaction: Act or process of liquefying or of rendering or becoming liquid; reduction to a liquid state.

liquid limit (LL): The minimum water mass content at which a small sample of soil will barely flow under a standard treatment. An Atterberg limit.

loading: The time rate at which material is applied to a treatment device involving length, volume, or other design factor.

loamy: Intermediate in texture and properties between fine-textured and coarse-textured soils. Includes all textural classes with the words loam or loamy as part of the class name.

loess: Material transported and deposited by wind, and consisting primarily of silt-sized particles.

lysimeter: (1) A device for measuring percolating and leaching losses from a column of soil under controlled conditions. (2) A device for measuring gains (precipitation and condensation) and losses (evapotranspiration) by a column of soil.

marl: Soft and unconsolidated calcium carbonate, usually mixed with varying amounts of clay and other impurities.

mesic: A soil temperature regime that has mean annual soil temperatures of 8°C or more but < 15°C and > 5°C difference between mean summer and mean winter soil temperatures at 50 cm below the surface.

mesophilic organism: An organism whose optimum temperature for growth falls in an intermediate range of approximately 15 - 35°C

metamorphic rock: Rock derived from pre-existing rocks, but that differs from them in physical, chemical, and mineralogical properties as a result of natural geological processes, principally heat and pressure, originating in the earth. The pre-existing rocks may have been igneous, sedimentary, or another form of metamorphic rock.

mica: A layer-structured aluminosilicate mineral group of the 2:1 type that is characterized by its high layer charge, which is usually satisfied by potassium.

microfauna: Protozoa, nematodes, and arthropods of microscopic size.

microflora: Bacteria (including actinomycetes), fungi, algae, and viruses.

mineralization: The conversion of an element from an organic form to an inorganic state as a result of microbial action.

mineral soil: A soil consisting predominately of, and having its properties determined predominantly by, mineral matter.

Mohr's circle of stress: A graphical representation of the components of stress acting across the various planes of a given point, drawn with reference to axes of normal stress and shear stress.

moisture content: The ratio (expressed as a percentage) of the weight of water in a given soil mass, to the weight of solid particles.

moisture-retention curve: A graph showing the soil moisture percentage (by weight) versus applied tension (or pressure). Points on the graph are usually obtained by increasing (or decreasing) the applied tension or pressure over a specified range.

Mollisols: An order in the U.S. system of soil taxonomy. Mineral soils that have a molic epipedon overlying mineral material with a base saturation of 50% or more when measured at pH 7. Mollisols may have an argillic, natric, albic, cambic, gypsic, calcic, or petrocalcic horizon, a histic epipedon, or a duripan, but not an oxic or spodic horizon.

mottles: Spots or blotches of different color or shades of color interspersed with the dominant color.

Munsell color system: A color designation system that specifies the relative degrees of the three simple variables of color: hue, value, and chroma.

neutral soil: A soil in which the surface layer, at least in the tillage zone, is in the pH range of 6.6 to 7.3.

nitrification: Biological oxidation of ammonium to nitrite and nitrate, or a biologically induced increase in the oxidation state of nitrogen.

nodule bacteria: The bacteria that fix di-nitrogen (N_2) within organized structures (nodules) on the roots, stems, or leaves of plants. Sometimes used as a synonym for "rhizobia."

normally consolidated soils: A soil deposit that has never been subjected to an effective pressure greater than the existing overburden pressure.

optimum moisture content: The water content at which a soil can be compacted to a maximum dry unit weight by a given compactive effort.

organic soil: A soil which contains a high percentage of organic matter through the soil zone.

oven-dry soil: Soil which has been dried at 105°C until it reaches constant weight.

overconsolidated soil deposit: A soil deposit that has been subjected to an effective pressure greater than the present overburden pressure.

oxidation state: The number of electrons to be added (or subtracted) from an atom in a combined state to convert it to the elemental form.

Oxisols: An order in the U.S. system of soil taxonomy. Mineral soils that have an oxic horizon within 2 m of the surface or plinthite as a continuous phase within 30 cm of the surface, and that do not have a sopdic or argillic horizon above the oxic zone.

pans: Horizons or layers, in soil that are strongly compacted, indurated, or very high in clay content.

parent material: The unconsolidated and more or less chemically weathered mineral or organic matter from which the solum of soil is developed by pedogenic processes.

particle density: The mass per unit volume of the soil particles. In technical work, usually expressed as grams per cubic centimeter.

particle size: The effective diameter of a particle measured by sedimentation, sieving, or micrometric methods.

particle size distribution: The amounts of the various soil separates in a soil sample, usually expressed as weight percentages.

ped: A unit of soil structure such as an aggregate, crumb, prism, block, or granule, formed by natural processes (in contrast to a clod, which is formed artificially).

pedon: A three-dimensional body of soil with lateral dimensions large enough to permit the study of horizon shapes and relations. Its area ranges from 1 to 10 square meters. Where horizons are intermittent or cyclic, and recur at linear intervals of 2 to 7 meters, the pedon includes one-half of the cycle. Where the cycle is less than 2 m, or all horizons are continuous and of uniform thickness, the pedon has an area of approximately one square meter. If the horizons are cyclic, but recur at intervals greater than 7 m, the pedon reverts to the 1 square meter size, and more than one soil will usually be represented in each cycle.

percent compaction: The ratio, expressed as a percentage, of dry unit weight of a soil to the maximum weight obtained in a laboratory compaction test.

percent saturation: The ratio, expressed as a percentage of the volume of water in a given soil to the total volume of intergranular space (voids).

percolation, soil water: The downward movement of water through soil; especially, the downward flow of water in saturated or nearly saturated soil at a hydraulic gradient of the order of 1.0 or less.

permeability, soil: The property of a porous medium itself that expresses the ease with which gases, liquids, or other substances can flow through it, and is the same as intrinsic permeability.

pH, soil: The negative logarithm of the hydrogen-ion activity of a soil. The degree of acidity (or alkalinity) of a soil as determined by means of a glass or other suitable electrode or indicator, at a specified moisture content or soil-water ratio, and expressed in terms of the pH scale.

physical weathering: The breakdown of rock and mineral particles into smaller particles by physical forces such as frost action.

piezometer: An instrument for measuring pressure head.

plastic limit (PL): The minimum water mass content at which a small sample of soil material can be deformed without rupture. An Atterberg limit.

plate count: A count of the number of colonies formed on a solid culture medium when inoculated with a small amount of soil. The technique has been used to estimate the number of certain organisms in the soil.

pore-size distribution: The volume fractions of the various size ranges of pores in a soil, expressed as a percentage of the soil bulk volume.

porosity: The volume of pores in a soil sample, divided by the bulk volume of the sample.

potash: Term used to refer to potassium or potassium fertilizers.

profile, soil: A vertical section of the soil through all its horizons and extending into the parent material.

reaction, soil: The degree of acidity or alkalinity of a soil, usually expressed as a pH value.

regolith: The unconsolidated mantle of weathered rock and soil material on the earth's surface; loose earth material above solid rock (similar to the term "soil" as used by many engineers).

saline soil: A nonsodic soil containing sufficient soluble salt to adversely affect the growth of most crop plants. The lower limit of saturation extract electrical conductivity of such soils is conventionally set at 0.4 semens per square meter.

sand: A soil particle between 0.05 and 2.0 mm in diameter. Commonly expressed as: very coarse sand, coarse sand, medium sand, fine sand, and very fine sand.

saprolite: Weathered rock materials that may be soil parent material.

saturate: (1) To fill all the voids between soil particles with a liquid; (2) to form the most concentrated solution possible under a given set of physical conditions in the presence of an excess of the solute; (3) to fill to capacity, as the adsorption complex with a cation species.

shear force: A force directed parallel to the surface element across which it acts.

silt: Soil particles between 0.05 and 0.002 in equivalent diameter.

shrinkage limit: The maximum water content at which a reduction in water content will not cause a decrease in volume of the soil mass.

soil: (1) The unconsolidated mineral or material on the immediate surface of the earth that serves as a natural medium for the growth of land plants; 2) the unconsolidated mineral or organic matter on the surface of the earth that has been subjected to and influenced by genetic and environmental factors of parent material, climate (including water and temperature effects), macro- and micro-organisms and topography, all acting over a period of time and producing a product—soil—that differs from the material from which it is derived in many physical, chemical, geological, and morphological properties and characteristics.

soil air: The soil atmosphere; the gaseous phase of the soil, being that volume not occupied by solid or liquid.

soil structure: The combination or arrangement of primary soil particles into secondary particles, units, or peds. These secondary units may be, but usually are not, arranged in the profile in such a manner as to give a distinctive, characteristic pattern. The secondary units are characterized and classified on the basis of size, shape, and degree of distinctness into classes, types, and grades, respectively.

soil suction: A measure of the force of water retention in unsaturated soil. Soil suction is equal to a force per unit area that must be exceeded by an externally applied suction to initiate water flow from the soil. Soil suction is expressed in standard pressure terms.

specific surface: The solid-particle surface area divided by the solid-particle mass or volume, expressed in m^2kg^{-1} or $m_2/m^3 = m^{-1}$, respectively.

Spodosols: An order in the U.S. system of soil taxonomy. Mineral soils that have a spodic horizon or a placic horizon that overlies a fragipan.

Stokes' Law: The equation expressing the force of viscous resistance on a smooth, rigid sphere moving in a viscous fluid, namely:

$$F = 3\pi\eta DV$$

where:

    F = viscous force resistance

    π = 3.1416

    η = the fluid viscosity

D = the diameter of the sphere

V = the velocity of fall or movement

strength: Maximum stress which a material can resist without failing for a given type of load.

subsoil: In general concept, that part of the soil below the depth of plowing.

tensile strength (unconfined or uniaxial tensile strength): The load per unit area at which an unconfined cylindrical specimen will fail in a simple tension (pull) test.

tensiometer: A device for measuring the negative pressure (or tension) of water in soil *in situ*; a porous, permeable ceramic cup connected through a tube to a manometer or vacuum gauge.

thermophile: An organism that grows readily at temperatures > 45°C.

tile drain: Concrete, plastic, or ceramic perforated and placed at suitable depth and spacings in the soil or subsoil to provide water outlets from the soil.

till (glacial): Unstratified glacial drift deposited by ice and consisting of clay, sand, gravel, and boulders, intermingled in any proportion.

tortuosity: The nonstraight nature of soil pores.

Ultisols: An order in the U.S. system of soil taxonomy. Mineral soils that have an argillic horizon with a base saturation of < 35% when measured at pH 8.2. Ultisols have a mean annual soil temperature of 8°C or higher.

udic: A soil moisture regime that is neither dry for as long as 90 cumulative days, nor for as long as 60 consecutive days in the 90 days following the summer solstice at periods when the soil temperature at 50 cm below the surface is above 5°C.

unsaturated flow: The movement of water in a soil which is not filled to capacity with water.

Vertisols: An order in the U.S. system of soil taxonomy. Mineral soils that have 30% or more clay, deep wide cracks when dry, and either gilgai micro-relief, intersecting slickensides, or wedgeshaped structural aggregates tilted at an angle from the horizon.

void ratio: The ratio of the volume of void space to the volume of solid particles in a given soil mass.

volumetric water content: The soil-water content expressed as volume of water per unit bulk volume of soil.

water tension (or pressure): The equivalent negative pressure in the soil water. It is equal to the equivalent pressure that must be applied to the soil water to bring it into hydraulic equilibrium through a porous permeable wall or membrane, with a pool of water with the same composition.

xeric: A soil moisture regime common to Mediterranean climates that have moist cool winters and warm dry summers. A limited amount of water is present but does not occur at optimum periods for plant growth. Irrigation or summer-fallow is commonly necessary for crop production.

zone of aeration: That part of the ground in which the voids are not continuously saturated.

Bibliography

Abdul, S.A., Migration of petroleum products through sandy hydrogeologic systems, *Ground Water Monitoring Rev.*, VIII, 4, 73, 1988.

Aber, J.D. and J.M. Mellilo, *Terrestrial Ecosystems*, W.B. Saunders Philadelphia, PA, 1991.

Atlas, R.M., *Basic and Practical Microbiology*, Macmillan Publishing Co., New York, 1986.

Atlas, R.M., Microbial degradation of petroleum hydrocarbons: an environmental perspective, *Microb. Rev.*, 45, 180, 1981.

Alexander, M., Microbial formation of environmental pollutants, *Adv. Appl. Microbiol.*, 18, 1, 1974, D. Perlman, Ed., Academic Press, New York.

American Petroleum Institute (1985): Detection of Hydrocarbons in Groundwater by Analysis of Shallow Soil Gas/Vapor; API Publication No. 4394, Washington, D.C.

American Society of Agronomy (1981, reprinted 1985): Chemistry in the Soil Environment; ASA Special Publication No. 40, Madison, WI.

American Society of Agromony (1984): Soil Testing:Correlating and Interpreting the Analytical Results; Madison WI.

American Society for Testing and Materials Committee D-18 (1979): Tentative Definitions of Terms and Symbols Relating To Soil Mechanics; ASTM D 653-42T, Annual Book of ASTM Standards, Part 19; ASTM, Philadelphia, PA.

American Society for Testing and Materials Committee D-18 (1992): Soil and Rock, Dimension Stone, Geosynthetics; Vol. 04.08 of the 1992 Annual Book of ASTM Standards, Philadelphia, PA.

Bowman, R.S. and D.P. Stevens, Field Study of Multidimensional Flow and Transport in the Vadose Zone, WRRI Report No. 262, New Mexico Water Resources Research Report, Las Cruces, NM, 1987.

Berger, K.C. *Sun, Soil and Survival: An Introduction to Soils*, University of Oklahoma Press, Norman, OK, 1978.

Brady, N.C., *The Nature and Properties of Soil*, 10th ed., Macmillan Publishing Co., New York, 1990.

Brewer, R., *Fabric and Material Analysis of Soils*, R.E. Krieger Publishing Co., Huntington, NY, 1976.

Broughton, W.J., Ed., *Nitrogen Fixation*, Vol.1, Clarendon Press, Oxford, 1981.

Bryan, A.H., C.A. Bryan, and C.B. Bryan, *Bacteriology: Principles and Practice*, 6th ed., College Outline Series, Barnes and Noble, New York, 1968.

Carter, V.G. and T. Dale, *Topsoil and Civilization*, rev. ed., University of Oklahoma Press, Norman, OK, 1981.

Childs, E.C., *Soil Moisture Theory*, Vol. 2, Advances in Hydroscience, V.T.Chow, Ed., Academic Press, New York, 73, 1967.

Das Braja, M., *Principles of Geotechnical Engineering*, PWS-Kent Publishing Co., Boston, MA, 1985.

Day, R.A. and A.L. Underwood, *Quantitative Analysis*, 5th ed., Prentice-Hall Co., Englewood Cliffs, NJ, 1986.

Dragun, J., The Soil Chemistry of Hazardous Materials, Hazardous Materials Control Research Institute, Silver Spring, MD, 1988.

Environmental Protection Agency, ChemFlo: One Dimensional Water and Chemical Movement in Unsaturated Soils, EPA/600/8-89/076, Robert S. Kerr Environmental Research Laboratory, ADA, OK, 1989.

Environmental Protection Agency, Cleanup of Releases from Petroleum UST's: Selected Technologies, EPA/530/UST-88/001, PB88-241856, 1988.

Environmental Protection Agency, Technology Transfer: Description and Sampling of Contaminated Soils: A Field Pocket Guide, EPA/625/12-91/002, 1991.

Environmental Protection Agency, Direct/Delayed Response Project: Quality Assurance Plan for Soil Sampling, Presentation and Analysis, EPA/600/8-87/021, Office of Acid Deposition, Environmental Monitoring and Quality Assurance, Washington, DC, 1987.

Environmental Protection Agency, Permit Guidance Manual on Unsaturated Zone Monitoring for Hazardous Waste Management Units, EPA/530/SW-86-040, Environmental Monitoring Systems Laboratory, Las Vegas, NV, 1986.

Environmental Protection Agency, The RETC Code for Quantifying the Hydraulic Functions of Unsaturated Soils, EPA/600/2-91/065, Office of Research and Development, Washington, DC, 1991.

Environmental Protection Agency, Soil Properties, Classification and Hydraulic Conductivity Testing, SW-925, Office of Solid Waste, Washington, DC, 1984.

Environmental Protection Agency, Soil Sampling Quality Assurance Users Guide, 2nd ed., EPA/600/8-89/046, Environmental Monitoring Systems Laboratory, Office of Research and Development, Las Vegas, NV, 1989.

Environmental Protection Agency, Seminar Publication: Transport and Fate of Contaminants in the Subsurface, EPA/625/4-89/019, Center for Environmental Research Information, Cincinnati, OH, 1989.

Fenchel, T. and T.H. Blackburn, *Bacteria and Mineral Cycling*, Academic Press, New York, 1979.

Fetter, C.W., *Applied Hydrogeology*, Charles E. Merrill Publishing Co., Columbus, OH, 1980.

Fetter, C.W., *Applied Hydrogeology*, 2nd ed., Macmillan Publishing Co., New York, 1988.

Fuller, W.H. and A.W. Warrick, *Soils in Waste Treatment and Utilization, Vol.1, Land Treatment*, CRC Press, Boca Raton, FL, 1985.

Gray, F. and M.H. Roozitalab, *Benchmark and Key Soils of Oklahoma: a Modern Classification System*, Oklahoma State University, Stillwater, (n.d.).

Hattori, T., *Microbial Life in the Soil, An Introduction*, Marcel Deckker, New York, 1973.

Hillel, D., *Introduction to Soil Physics*, Academic Press, San Diego, CA, 1980.

Hillel, D., *Soil and Water*, Academic Press, New York, 1971.

Hinchee, R.E., D.C. Downey, and E.J. Coleman, Enhanced bioreclamation, soil venting and groundwater extraction: a cost effectiveness and feasibility comparison, in Proc. of the NWWA and API Conference on Petroleum Hydrocarbons and Organic Chemicals in Groundwater—Prevention, Detection and Restoration, November 1987, 147.

Jensen, R.L., *Statistical/Survey Techniques*, John Wiley & Sons, New York, 1978.

Jury, W.A., W.R. Gardner and W.H. Gardner, *Soil Physics*, 5th ed., John Wiley & Sons, New York, 1991.

Kostecki, P.T. and E.J. Calabrese, *Petroleum Contaminated Soils*, Vol. 1, Lewis Publishers, Chelsea, MI, 1989.

Lambe, T.W., *The Nature and Engineering Significance of the Structure of Compacted Clay*, MIT Press, Cambridge, MA, 1958.

Laskin, A.I., Ed., *Advances in Microbiology*, Academic Press, New York, 1982.

Lindburg, M.R., *Civil Engineers Reference Manual*, 5th ed., Professional Publications, Inc., Belmont, CA, 1989.

Loughnan, F.C., *Chemical Weathering of the Silicate Minerals*, American Elsevier Publishing Company, New York, 1969.

Lyon, T.L. and H.O. Buchman, *The Nature and Properties of Soil*, Macmillan Company, New York, 1937.

Michigan Department of Highways, Field Manual of Soil Engineering, 5th ed., Lansing, MI, 1970.

Overcash, M.R. and D. Pal, *Design of Land Treatment Systems for Industrial Wastes - Theory and Practice*, Ann Arbor Science Publishers, Ann Arbor, MI, 1979.

Page, A.L., R.H. Miller, and D.R. Keeney, Eds., *Methods of Soil Analysis*, 2nd ed., Chemical and Microbiological Properties, American Society of Agronomy, No. 9, Part 2, Madison, WI, 1982.

Parcher, J.V. and R.E. Means, *Soil Mechanics and Foundations*, Charles Merrill Company, Columbus, OH, 1968.

Ripple, C.D., J. Rubin, and T.E.A. Van Hylckama, Estimating Steady-State Evaporation Rates from Bare Soil under Conditions of High Water Table, U.S. Geological Survey Water Supply Paper 2019-A, 1972.

Sawhney, B.L. and K. Brown, Eds., Reactions and Movement of Organic Chemicals in Soils, Special Publication No. 27, Soil Science Society of America, Madison, WI, 1989.

Schilfgaarde, J. Van, Ed., Drainage for Agriculture, No. 17 in the Series, Agronomy, American Society of Agronomy, Madison, WI, 1974.

Sistrom, W.R., Microbial Life - from the Modern Biology Series, Holt, Rinehart and Winston, New York, 1962.

Soil Science Society of America, Chemical Mobility and Reactivity in Soil Systems, SSSA Special Publication No. 11, Madison, WI, 1983, 3rd printing 1992.

Soil Science Society of America, Glossary of Terms, Soil Science Society of America, Madison, WI, 1975.

Soil Science Society of America, Water Potential Relations in Soil Microbiology, Special SSSA Publication No. 9, Madison, WI, 1981.

Solomons, T.W.G. *Organic Chemistry*, 4th ed., John Wiley & Sons, New York, 1988.

Spangler, M.G. and R.L. Handy, *Soil Engineering*, 3rd ed., Harper and Row, New York, 1973.

Sparks, D.L., *Kinetics of Soil Chemical Processes*, Academic Press, San Diego, CA, 1989.

Stevens, D.B., E. Hicks, and T. Stein, Field Analysis on the Role of Three-Dimensional Moisture Flow in Ground-Water Recharge and Evaporation, Technical Completion

Report No. 260, New Mexico Water Resources Institute, Las Cruces, NM (undated).

Stolp, H., *Microbial Ecology: Organisms, Habitats, Activities*, Cambridge University Press, Cambridge, UK, 1988.

Swartzendruber, D., The flow of water in unsaturated soils, in *Flow Through Porous Media*, R.J.M. Deweist, Ed., Academic Press, New York, 215, 1969.

Tarbuck, E.J. and F.K. Lutgens, *The Earth, An Introduction to Physical Geology*, Charles E. Merrill Publishing Company, Columbus, OH, 1984.

Terzaghi, K. and R. Peck, *Soil Mechanics in Engineering Practice*, 2nd ed., John Wiley & Sons, New York, 1968.

Testa, S.M., *Geological Aspects of Hazardous Waste Management*, Lewis Publishers, Boca Raton, FL, 1993.

Testa, S.M. and D.L. Winegardner, *Restoration of Petroleum-Contaminated Aquifers*, Lewis Publishers, Boca Raton, FL, 1991.

U.S. Department of Agriculture, Water: Year Book of Agriculture, 1955, U.S. Government Printing Office, Washington, DC, 1955.

U.S. Department of the Interior, Earth Manual, 2nd ed., A Water Resources Technical Publication, U.S. Government Printing Office, Washington, DC, 1974.

U.S. Soil Conservation Service, Glossary of Selected Geologic and Geomorphic Terms, Western Technical Service Center, Portland, OR, 1977.

U.S. Soil Conservation Service, National Engineering Handbook, U.S. Government Printing Office, Washington, DC, 1972.

Walker, W.H., Where have all the toxic chemicals gone?, *in Ground Water*, 11, 2, 11, 1973.

Wagner, R.E., Ed., Guide to Environmental Analytical Methods, Genium Publishing Corporation, Schenectady, NY, 1992.

Watson, K.K., A recording field tensiometer with rapid response characteristics, in *J. Hydrol.*, 5, 33, 1967.

Index

ERRATA

Page 7 — Table 2.1 First column: Second Coarse Gravel should read: Course Sand

Page 33 — Line 10 from the top: The second *monoclinic* should be deleted

Page 53 — Line 10 from the bottom: There should be a comma following *blasting*

Page 59 — Equation should read: Volume of Solids $= \dfrac{1055\text{gm}}{2.70\text{gm/cm}^3} = 390.7\ \text{cm}^3$

Page 77 — Bottom formula should read: $W = \dfrac{\text{weight of water}}{\text{dry weight of soil}}$

Page 86 — Second line from bottom should read: Hydraulic Gradient (I) = .0005

Page 87 — Bottom formula should read: $K = k\delta g/\mu$

Page 102 — Section 6.3.2 should read: 6.4.2

Page 104 — Line 12: Mussel should read *Munsel*

Page 108 — Lines 9 and 10 from the top: OH_ should read: OH^-

Page 110 — Section 6.9.1: $NO_{3\_}$ should read: NO_3^-

Page 119 — Line 10 from the top: There is a comma following *WC*

Page 121 — Henry's law constant should read: $K_h =$

Page 134 — Line 5 from the bottom: There should be a closed parethesis following *Max*)

Page 140 — Section 7.9.1: The formula should end with: $4NH_4^+$

Page 142 — Section 7.9.5: $NO_{3\_}$ should read: NO_3^-

Page 143 — Section 7.10.1: The formula should begin: SO_4^{-2}

Page 226 — Line 14 from the bottom: cliche should read: caliche

Page 235 — Under Darcy's Law, it should read:
   H = the height of the water on top of the bed, and
   k = a coeffcient...

Milton Keynes UK
Ingram Content Group UK Ltd.
UKHW040444071024
449327UK00020B/985

9 780367 448851